PRECALCULUS EXPERIMENTS
WITH THE CASIO GRAPHICS CALCULATOR

Lawrence G. Gilligan
University of Cincinnati

D. C. Heath and Company
Lexington, Massachusetts   Toronto

The figures on pages 149 and 150 appear courtesy of Casio, Inc.

Published simultaneously in Canada.

Printed in the United States of America.

International Standard Book Number: 0-669-27644-8

1 2 3 4 5 6 7 8 9 0

PRECALCULUS EXPERIMENTS WITH THE CASIO GRAPHICS CALCULATOR

CONTENTS

Preface

This manual is part of a ten-year love affair between the author and the use of technology in mathematics instruction. While computers and software are progressing at a rapid pace, there probably is no better tool, in terms of providing students with concepts and accessibility, than the graphics calculators of the 1990's. It is my goal that this manual will become a bridge in the students' learning process -- a bridge between 1) the mathematics taught in the classroom and text-book and 2) the capabilities of the graphing calculator. It is meant to replace neither but to enhance each.

The manual includes 21 experiments designed to take between one and two hours with one of the family of Casio graphics calculators. (A TI-81 version of this manual is also available.) The author prefers the laboratory experience where students work in pairs but each submit a separate report. This encourages brainstorming and creativity yet also allows for individual differences and efforts. These reports are graded and weighted collectively as the approximate equivalent of an hourly examination grade. Of course, individual instructors will use the the experiments as they best fit their course.

There are over one hundred graphics screens reproduced in this manual to help readers conceptualize the topics of precalculus. These screens are meant to be virtual "snapshots" of the resulting Casio screens and include a border and grid-like dot pattern for additional clarity.

The manual is textbook-independent and was written to correlate with any of a variety of popular precalculus and college algebra/trigonometry textbooks on the market today. Comparing the table of contents of the manual with that of the textbook is the obvious way to determine the best time to introduce a particular experiment, but we also provide the following comparison of the manual's topics with the section numbers of two of the most popular textbooks.

Topic Comparison Guide

Manual Contents	Precalculus[1] L&H	Algebra and Trig[2] L&H
#1: Graphing Straight Lines	2.3	3.3
#2: Graphs of Functions	2.5	3.5
#3: Inverse Functions	2.7	3.7
#4: Quadratic Functions	3.1	4.1
#5: Polynomial Functions of Higher Degree	3.2	4.2
#6: Rational Functions	3.7	5.1
#7: Exponential Functions	4.1	6.1
#8: Logarithmic Functions	4.2	6.2
#9: Approximating Solutions to Exponential and Logarithmic Equations	4.4	6.4
#10: Introduction to Trigonometry	5.1	7.1
#11: Evaluating Trigonometric Functions	5.2, 5.3, 5.4	7.2, 7.3
#12: Graphs of Sine and Cosine	5.5	7.4
#13: Graphs of Other Trigonometric Functions	5.6	7.5
#14: Inverse Trigonometric Functions	5.8	7.7
#15: Approximating Solutions to Trigonometric Equations	6.3	8.3
#16: Solving Systems of Equations Graphically	8.1, 8.2	10.1, 10.2
#17: Determinants and Cramer's Rule	9.4, 9.5, 9.6	11.4, 11.5, 11.6
#18: Conic Sections Part I: Circles and Parabolas	11.1	5.3
#19: Conic Sections Part II: Ellipses	11.2	5.3, 5.4
#20: Conic Sections Part III: Hyperbolas	11.3	5.3, 5.4
#21: Polar Coordinates	11.5, 11.6	-

---

[1]*Precalculus, Second Edition* by Roland E. Larson and Robert P. Hostetler (D. C. Heath and Company, Lexington, MA: 1989)

[2]*Algebra and Trigonometry, Second Edition* by Roland E. Larson and Robert P. Hostetler (D. C. Heath and Company, Lexington, MA: 1989)

Acknowledgments

This manual never would have been possible without the diligence and enthusiasm shown by Ann Marie Jones, Mathematics Editor at D. C. Heath and Company. Her direction, support, and concern for the project were overwhelming and greatly appreciated. Also, Carolyn Johnson, Editorial Associate, worked on this and several other of my projects with D. C. Heath and her organizational skills and smiles have always been valued.

The manuscript went through several revisions and scores of extremely valuable suggestions came from the following reviewers: Professor Grace Cascio, Northeast Louisiana State University; Professor Margaret Friar, Grand Valley State University; Professor William Hemme, St. Petersburg Junior College - Clearwater; and Professor Lee Stiff, North Carolina State University. Thank you all.

Finally but foremost, for my family (Sue, Andy, and Katie) who provided encouragement and understanding for the long hours this project separated me from them, I love you. It is to you that I dedicate this work.

... for the students ...

In the past, some of the best suggestions and criticisms for this type of manual come from the students who use it. If, at anytime in your experience with this text, you care to offer any criticism, or you have any questions, or you just want to make a comment, please feel free to contact me. I promise a reply.

Professor Lawrence G. Gilligan
Department of Mathematics, Physics and Computing Technology
OMI College of Applied Science
University of Cincinnati
2220 Victory Parkway
Cincinnati, OH 45206-2822
(513) 556-4868

Other Manuals by the Author

Precalculus Experiments with the TI-81 Graphics Calculator by Lawrence G. Gilligan (D. C. Heath and Company, Lexington, MA: 1991)

Calculus and the DERIVE Program: Experiments with the Computer, Second Edition by Lawrence G. Gilligan and James F. Marquardt, Sr. (Gilmar Publishing, P.O. Box 6376, Cincinnati, OH 45206: 1991)

Why a *Graphics* Calculator?

Undoubtedly, by now you have used a calculator, probably even a *scientific* calculator to perform arithmetic operations and you can appreciate the speed and accuracy such a device has. Certainly, a calculator can help us save time in our mathematics, science, business and engineering courses. If you are saving time, how are you going to spend that newly found time? Hopefully, at least in the eyes of your mathematics instructor, that freed-up time will be spent by you learning more mathematics *concepts* and that is where a graphics calculator can really help!

For example, suppose you needed to know what the greatest value an expression like $x + 2x^2 - 3x^4$ could be. In "the old days" (that is, *before* graphics calculators), you may have experimented by trying many values of x and gotten frustrated (and maybe even a sore index finger!) when entering all those different x values. But, the *concept* of achieving a maximum is best understood by looking at a picture (or graph). You will find how to use the Casio graphics calculator to quickly and visually supply you with an approximation to the problem. In Figure 1 below, we see that $x + 2x^2 - 3x^4$ has a greatest value of about 0.96 and that occurs when $x \approx 0.7$.

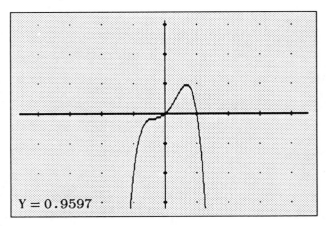

We will see later in this manual that a graph of $y = x + 2x^2 - 3x^4$ is a pictorial way of examining values of $x + 2x^2 - 3x^4$. The highest point plotted corresponds with the greatest value of the expression. In this case, that happens at about $x = 0.7$.

Figure 1.

So, the advantage of the Casio to quickly and accurately plot a very large number of points, will be emphasized in this manual as will the ability to store and run programs (series of statements in logical sequence).

Numerical Computation on the Casio Graphics Calculator

The information in this manual applies to four models of Casio graphics calculators: the fx-7000G, the fx-7500G, the fx-8000G and the fx-8500G. Specific arrangement of keys and the different storage capacity (for programming) are the major differences. Two keys are of particular interest, the SHIFT key and the ALPHA key. These keys, like the shift key on a typewriter, are used in conjunction with other keys to increase each key's functionality. The SHIFT key is a different color (blue on the 7500G and gold on the others) and it will access the colored feature on the key pressed after it is pressed. Similarly, the ALPHA key is used with other keys to access the (upper-case) 26 letters of the alphabet and some additional special characters.

Another key, the MODE key needs mentioning here. For the calculations we are about to do, enter MODE 7. When the word "Fix" comes on the screen enter 5 EXE. This means that answers will be displayed to 5 decimal places of precision. The MODE key also allows you to choose scientific notation (MODE 8) and whether you want to work with angles in degrees (MODE 4) or radians (MODE 5). Let's make sure that you are in the right mode settings for the discussions in this introduction by displaying the mode: press the M Disp key. You should see:

```
        * * * * MODE * * * *
sys mode    :   RUN
cal mode    :   COMP
  angle     :   Rad
  display   :   Fix 5

        Step      0
```

Let's begin computing by attempting to find the value of $-8 + 12 - (-2)$. Notice the following two keys on the calculator: $(-)$ and $-$. They are considerably different! To enter negative numbers, we use the $(-)$ key; for subtraction, we use the $-$ key. Thus, to find the value of $-8 + 12 - (-2)$, we would enter the following keystrokes:

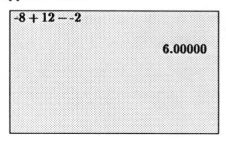

$\boxed{(-)}\; 8 \;\boxed{+}\; 1\;2 \;\boxed{-}\; \boxed{(-)}\; 2 \;\boxed{\text{EXE}}$ [1]

The screen display appears below:

```
-8 + 12 - -2

                  6.00000



```

For a more complicated calculation, like $\sqrt{6^2 - 3^2}$, we need to make use of the left parenthesis key, $\boxed{(}$, and the right parenthesis key, $\boxed{)}$. The keystrokes necessary to enter are:

$\boxed{\sqrt{}}\; \boxed{(}\; 6 \;\boxed{x^2}\; \boxed{-}\; 3 \;\boxed{x^2}\; \boxed{)}\; \boxed{\text{EXE}}$

Now, the screen should resemble:

```
-8 + 12 - -2
                  6.00000
√(6² - 3²)

                  5.19615

```

Suppose you erred and really wanted to calculate $\sqrt{6^2 + 3^2}$. There is no need to re-enter the new expression; instead, we can *edit* the previously entered expression. To do this, press the left arrow editing key, $\boxed{\Leftarrow}$ on the *fx-7000G* and $\boxed{\triangleleft}$ on the *fx-7500G*. Now, the answer disappears and the expression reappears. Press the left arrow key four more times until the cursor is under the "$-$" symbol and enter the corrected key, $\boxed{+}$ followed by the $\boxed{\text{EXE}}$ key. The screen display should look like this:

```
√(6² + 3²)
                  6.70820



```

_____

[1]For the sake of comparison for those with other scientific algebraic logic calculators, we mention here that the $\boxed{(-)}$ key corresponds to a $\boxed{+/-}$ or $\boxed{\text{Ch Sign}}$ key on those calculators and the $\boxed{\text{EXE}}$ key corresponds to the $\boxed{=}$ key for computations.

It should be noted that the most previously computed value, in this case 6.70820, is now stored in the "Answer" memory and is accessible by pressing the $\boxed{\text{Ans}}$ key. Thus, to find $(6.70820)^2$, we merely have to enter $\boxed{\text{Ans}}$ $\boxed{x^2}$:

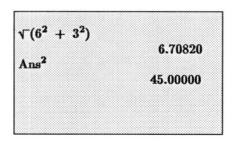

Exponents and roots are calculated using the $\boxed{x^y}$ and the $\boxed{^x\sqrt{}}$ keys, respectively. Several examples follow showing the exact keystrokes to enter to obtain the specified answer. The reader should verify these answers.

$$\frac{4}{\sqrt{11} - \sqrt{7}}$$

$4 \boxed{\div} \boxed{(} \boxed{\sqrt{}} 11 \boxed{-} \boxed{\sqrt{}} 7 \boxed{)} \boxed{\text{EXE}}$

Ans: 5.96238

$$\sqrt[5]{72^2} \quad \text{or} \quad 72^{2/5}$$

$5 \boxed{^x\sqrt{}} \boxed{(} 72 \boxed{x^2} \boxed{)} \boxed{\text{EXE}}$

Ans: 5.53265

$$\frac{-3 - \sqrt{3^2 - 4(2)(-3)}}{2 \cdot 2}$$

$\boxed{(} \boxed{(-)} 3 \boxed{-} \boxed{\sqrt{}} \boxed{(} 3 \boxed{x^2} \boxed{-} 4 \boxed{\times} 2 \boxed{\times} \boxed{(-)} 3 \boxed{)} \boxed{)} \boxed{\div} \boxed{(} 2 \boxed{\times} 2 \boxed{)} \boxed{\text{EXE}}$

Ans: -2.18614

The next two calculations involve the number pi.[2]

$$\frac{\pi + 2}{\sqrt{3}}$$

$\boxed{(} \boxed{\text{SHIFT}} \boxed{\pi} \boxed{+} 2 \boxed{)} \boxed{\div} \boxed{\sqrt{}} 3 \boxed{\text{EXE}}$ Ans: 2.96850

---

[2]An important notation convention for using this manual needs some clarification here. To access the constant π, we need to SHIFT the EXP key. We choose to write that as $\boxed{\text{SHIFT}}$ $\boxed{\pi}$ rather than $\boxed{\text{SHIFT}}$ $\boxed{\text{EXP}}$ to emphasize that we are selecting π. So, to access the alphabetic character "A", for example, we will write $\boxed{\text{ALPHA}}$ $\boxed{\text{A}}$ as opposed to $\boxed{\text{ALPHA}}$ $\boxed{x^{-1}}$. Also, we do not place numbers in boxes and it is assumed when we write 3.78, for example, that the four obvious keys are pressed: $\boxed{3}$ $\boxed{.}$ $\boxed{7}$ $\boxed{8}$.

$\left| \dfrac{6 - \pi}{\sqrt[3]{7}} \right|$ $\boxed{\text{SHIFT}}$ $\boxed{\text{Abs}}$ $\boxed{(}$ $\boxed{(}$ 6 $\boxed{-}$ $\boxed{\text{SHIFT}}$ $\boxed{\pi}$ $\boxed{)}$ $\boxed{\div}$ $\boxed{\text{SHIFT}}$ $\boxed{\sqrt[3]{}}$ 7 $\boxed{)}$ $\boxed{\text{EXE}}$

Ans: 1.49426

We conclude this section on computation by mentioning that numbers may be entered in scientific notation directly by using the $\boxed{\text{EXP}}$ key. For example, to enter the number 2.67×10^4, enter 2.67 $\boxed{\text{EXP}}$ 4 $\boxed{\text{EXE}}$. Assuming the calculator is in MODE 7 with 5 fixed decimal points, the display should read:

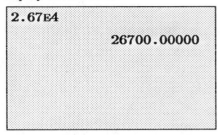

An alternate mode setting of $\boxed{\text{MODE}}$ 8 (for scientific notation) followed by the number of digits of precision, say 5, and then $\boxed{\text{EXE}}$ would show the following:

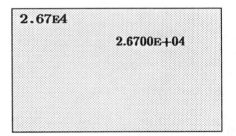

Finally, a number can be converted to *engineering notation* (where the power of 10 is always a multiple of 3) by pressing the $\boxed{\text{ENG}}$ key.

Data Storage (... or "Thanks for the Memory"!)

If a particular number needs to be "remembered" or is going to be used in many subsequent calculations, it can be stored by using the $\boxed{\rightarrow}$ ("assign to memory") key in conjunction with a memory name (the letters A through Z). For example, to store the number -19.66 in memory C, we enter the following keystrokes:

$\boxed{(-)}$ 19.66 $\boxed{\rightarrow}$ $\boxed{\text{ALPHA}}$ $\boxed{\text{C}}$ $\boxed{\text{EXE}}$

To check the contents of memory D, for instance, we enter $\boxed{\text{ALPHA}}$ $\boxed{\text{D}}$ $\boxed{\text{EXE}}$ and to increase the value in memory F by 12.5, we would enter:

$$\boxed{\text{ALPHA}}\ \boxed{\text{F}}\ \boxed{+}\ 12.5\ \boxed{\Rightarrow}\ \boxed{\text{ALPHA}}\ \boxed{\text{F}}\ \boxed{\text{EXE}}$$

To "zero" (or clear) the contents of a memory, we store the value of 0 in it. For example, to insure that the contents of memory T holds nothing, we enter the keystrokes

$$0\ \boxed{\Rightarrow}\ \boxed{\text{ALPHA}}\ \boxed{\text{T}}\ \boxed{\text{EXE}}$$

While it is generally not advised to clear *all* memories, this is accomplished by pressing the key combination $\boxed{\text{SHIFT}}\ \boxed{\text{Mcl}}\ \boxed{\text{EXE}}$.

Finally, we mention the use of the colon key, $\boxed{:}$. This key is used to enter more than one calculation or direction and only uses the $\boxed{\text{EXE}}$ key once. For example, suppose we need to calculate both the circumference (memory C) and area (memory A) of a circle of radius 5.25 cm (memory R). Compare the keystrokes with the screen display and note that we can enter quite a bit of information by incorporating the colon key:

Keystrokes

$5.25\ \boxed{\Rightarrow}\ \boxed{\text{ALPHA}}\ \boxed{\text{R}}\ \boxed{:}\ 2\ \boxed{\text{SHIFT}}\ \boxed{\pi}\ \boxed{\text{ALPHA}}\ \boxed{\text{R}}\ \boxed{\Rightarrow}\ \boxed{\text{ALPHA}}\ \boxed{\text{C}}\ \boxed{:}\ \boxed{\text{SHIFT}}\ \boxed{\pi}\ \boxed{\text{ALPHA}}\ \boxed{\text{R}}\ \boxed{x^2}$

$\boxed{\Rightarrow}\ \boxed{\text{ALPHA}}\ \boxed{\text{A}}\ \boxed{\text{EXE}}$

$\boxed{\text{ALPHA}}\ \boxed{\text{C}}\ \boxed{\text{EXE}}$

Display

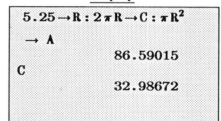

$$5.25 \rightarrow R : 2\pi R \rightarrow C : \pi R^2$$
$$\rightarrow A$$
$$\qquad\qquad 86.59015$$
$$C$$
$$\qquad\qquad 32.98672$$

← After $\boxed{\text{EXE}}$ is pressed, the value of A, the area is displayed.

← After $\boxed{\text{EXE}}$ is pressed a second time, the value of the circumference, C, is displayed.

The Graphing Keys

As we said earlier, what makes this calculator so powerful is its ability to graph points. Before we have the Casio execute a graph, we first must observe a very important graphing-related key, the $\boxed{\text{Range}}$ key. When this key is pressed, six values are displayed: Xmin (the smallest possible x value that can be plotted), Xmax (the largest possible x value), Xscl (the distance between hatch marks on the x-axis), Ymin (the smallest possible y value), Ymax (the largest possible y value), and Yscl (the distance between hatch marks on the y-

axis). The <u>default</u> range values are listed below:[3]

```
Range
Xmin:    -4.7
 max:    4.7
 scl:    1.
Ymin:    -3.1
 max:    3.1
 scl:    1.
```

Figure 2. The *default* range settings.

An important observation is that the graphing screen (and the text screen, too) is wider than it is high. For that reason, there are more plottable "points" (actually called *pixels*) horizontally than vertically. In fact, the ratio of width to height is about 3:2 or 1.5:1. For plotting situations where you want about the same distance between x-axis hatch marks as between y-axis hatch marks, the difference between Xmax and Xmin should be about 1.5 times the difference between Ymax and Ymin when both scales (Xscl and Yscl) equal 1.

To graph $y = x^3 - 4x^2$ using the default range settings, simply enter the following keystrokes:

$\boxed{\text{Graph}}\ \boxed{\text{ALPHA}}\ \boxed{\text{X}}\ \boxed{x^y}\ 3\ \boxed{-}\ 4\ \boxed{\text{ALPHA}}\ \boxed{\text{X}}\ \boxed{x^2}\ \boxed{\text{EXE}}$

Notice, in Figure 3, that we do not get a complete picture of the graph. That is why in Figures 4 and 5, we present two other variations. The reader is urged to try to duplicate each graphics screen by manually entering the range values (be sure to hit the $\boxed{\text{EXE}}$ key after each entry).

```
Range
Xmin:    -4.7
 max:    4.7
 scl:    1.
Ymin:    -3.1
 max:    3.1
 scl:    1.
```

With the default range settings above, the graph does not appear "complete." Notice that the ratio of X distance to Y distance is very close to 1.5 to 1:
$(4.7 - (-4.7)) : (3.1 - (-3.1))$

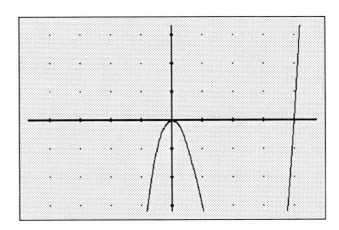

Figure 3.

```
Range

Xmin:    -8.
 max:    13.
 scl:     2.
Ymin:   -11.
 max:     3.
 scl:     2.
```

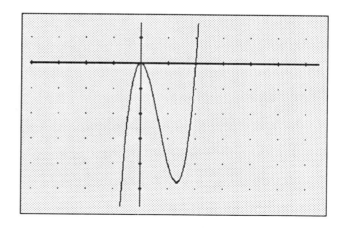

Notice that each hatch mark now
represents 2 units (Xscl = Yscl = 2).
The graph also appears a bit more
complete.

Figure 4.[4]

```
Range

Xmin:    -4.
 max:    10.
 scl:     2.
Ymin:   -11.
 max:    10.
 scl:     3.
```

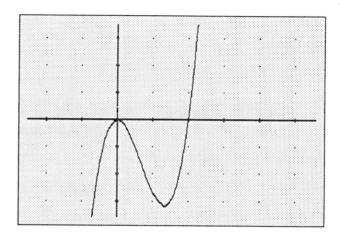

The x and y scales are now different;
the graph displays some distortion
but we get to "see" more of it!

Figure 5.

To clear the graphics screen type [Cls] [EXE].

As we progress through the labs in this manual, more details about the graphing
screen and related keys will be made. For now, suffice it to say that the range settings play a
vital role in graph-viewing and that we never graph a function without referring to the
[Range].

---

[4]To change the range values, press [Range] to display them and then enter each value followed by
the [EXE] key. To obtain the values shown in Figure 4, for example, press the following keys:
[(−)] 8 [EXE] 13 [EXE] 2 [EXE] [(−)] 11 [EXE] 3 [EXE] 2 [EXE]. Then, press [EXE] one more time to
redraw the graph with the new settings.

Although the actual program-writing will take place as we progress in this manual, suffice it to say here that there are three important MODES regarding programs. In RUN mode ($\boxed{\text{MODE}}$ 1 and the default mode) we can run stored programs by using the $\boxed{\text{PROG}}$ key and then entering one of the digits, 0 through 9 followed by $\boxed{\text{EXE}}$. To write a program, we first choose $\boxed{\text{MODE}}$ 2 and then, using the cursor movement keys, choose the number of the program we will write (or edit). The screen below appears:

```
sys mode :    WRT
cal mode :    COMP
   angle:     Rad
 display:     Fix 5

 4006 Bytes Free

Prog 0 1 2 3 4 5 6 7 8 9
```

← $\boxed{\text{MODE}}$ 5 chosen
← $\boxed{\text{MODE}}$ 7 then 5 $\boxed{\text{EXE}}$ entered

← Varies depending on programs in memory and the model of Casio.
← As programs are entered, number is replaced with an underscore, _.

To leave the program write/edit mode, enter $\boxed{\text{MODE}}$ 1, the run and calculation mode. Finally, to erase (or *clear*) a program in memory, $\boxed{\text{MODE}}$ 3 is chosen. (The screen is displayed below.) Using the cursor keys to select the number of the program to be erased, press the $\boxed{\text{AC}}$ key.

```
sys mode :    PCL
cal mode :    COMP
   angle:     Rad
 display:     Fix 5

 3215 Bytes Free

Prog 0 1 2 _ 4 5 6 7 8 9
```

NOTES

4. A series of steps is usually preceded by a label $\boxed{\text{LBL}}$. The series of steps involves plotting the point (x, y), incrementing y by some small value, say 0.5, and then plotting again until we have reached y's maximum value of 8.

4. LBL 1 : Plot X, Y :
   Y + .5 → Y : Y ≤ 8 ⇒ Goto 1 ◀

5. A "trick" for leaving the graphing window displayed (as opposed to the text window) is to graph $y = 0$.

5. Graph Y = 0.

Run the program and plot the vertical line $x = 2.5$. To return to the text window after program execution, press the $\boxed{\text{G}\longleftrightarrow\text{T}}$ key. A facsimile appears in Figure 1.3 below.

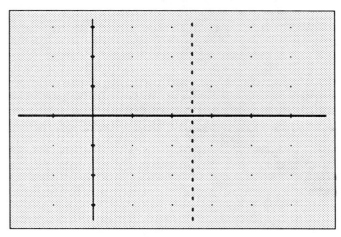

FIGURE 1.3. $x = 2.5$

NOTES

Name _____

Date _____

DIRECTIONS: Show all your work right on these pages. In Exercises 1 through 5 sketch the straight line graphs on the axes provided. Be sure to clear the graphics screen after each exercise via [Cls] [EXE].

1. $y = 2x - 2$

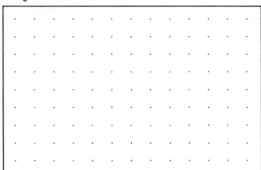

2. $y = -3x + 1$

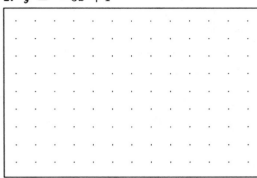

3. $y = \frac{1}{2}x + \frac{3}{2}$

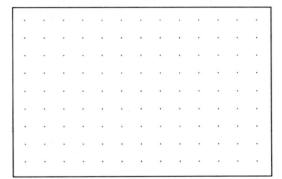

4. $2x - y = 4$ [Hint: first solve for y.]

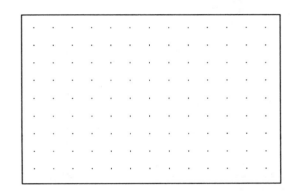

5. Graph each pair of equations in the grid provided.

a. $y = 2x$ and $y = -\frac{1}{2}x$

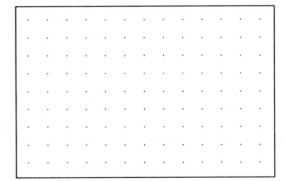

b. $y = -4x + 1$ and $y = \frac{1}{4}x + 1$

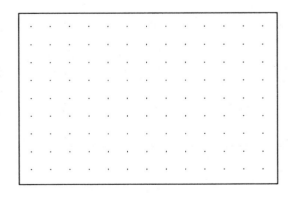

c. What observation do you have about each pair of lines? What in the equations makes it so?

In Exercises 6 through 8, use the program from Procedure #2 (program P0) to find the slope of the line segment connecting each of the following pairs of points:

6. $(4, 5)$ and $(-2, 11)$ Slope = _____

7. $(-2, 5)$ and $(3, -4)$ Slope = _____

8. $(3, -7)$ and $(3, 0)$ Slope = _____

9. Adapt the program in Procedure #2 to graph the line passing through the two entered points. (Assume the line is not vertical.) Here are some things to think about:

       1) Store the value of $(D - B) \div (C - A)$ in storage register M

       2) Calculate the y-intercept by computing $y_1 - x_1 \cdot \dfrac{y_2 - y_1}{x_2 - x_1}$. Store this in storage register G.

       3) Graph $Y = MX + G$

10. Use the program in Procedure #4 (program P1) to plot each of the following lines:

a) $x = 3$

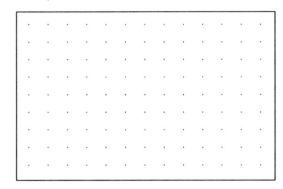

b) $x = -1.5$

c) $3x = 8$

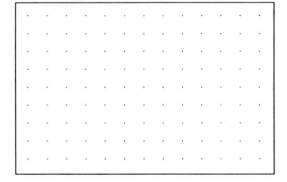

d) $2x - 1 = -5$

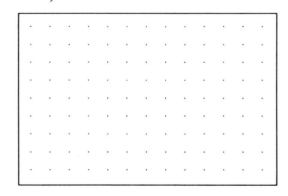

Graphs of Functions

INTRODUCTION

1. Your textbook provides you with a working definition of function, domain and range. Here, we will be interested in domain and range from the graphics calculator's viewpoint. This will mean entering the domain (the smallest and largest x-values for the interval of concern) using Casio's [Range] key. Then, we can determine the range of y values from inspecting the graph.

2. An odd function is one that is symmetric with respect to the origin; an even function is symmetric with respect to the y-axis. The graphics calculator can help us determine a function's parity (evenness or oddness) by examining the function's graph.

3. If $y = f(x)$ has a particular graph, then $y = f(x + c)$ represents a shift in the original graph by a factor of c units. The shift is to the left if c is positive and to the right if c is negative. Similarly, $y = f(x) + c$ represents a vertical shift in the original graph c units. If c is positive, the shift is upward; if c is negative, the shift is downward.

4. One of the most important functions in the early study of calculus is the absolute value function. We will use the Casio's graphing capabilities to examine $y = |x|$ and variations of that function.

PROCEDURES

Procedure 1. We have already used the [Range] key. Recall that it is composed of six settings: the maximum and minimum x-values, the x scale factor, the maximum and minimum y-values, and the y scale factor. Let's graph $y = 3x - x^2$ using the default range settings. (If you do not have the default range settings active, they can be obtained by pressing [SHIFT] [DEL] while in the range screen.) A graph similar to the one in Figure 2.1 appears.

Suppose that we are interested in finding the range for this function for $-1 \leq x \leq 3$. Notice by using the [Trace] key and then using the arrows, we can move the flashing pixel until we have located the maximum y value. The x-value of the maximum point is displayed; press the [SHIFT] [X↔Y] key and the y-value is displayed.

<u>Procedure 3.</u> Let us compare a function $y = f(x)$ with a variation of an added constant: $y = f(x+c)$. For example, consider two variations of the function $f(x) = x^2$: $y = (x-1)^2$ and $y = (x+2)^2$. So, the displacement constant, c, is -1 for one variation and $+2$ for the other. The three functions can be graphed on one set of the Casio's axes as depicted in Figure 2.8 below.

```
Range
Xmin:  -4.7
  max:   4.7
  scl:   1.
Ymin:  -0.5
  max:   5.
  scl:   2.
```

Note: If $c = -1$, the original graph is shifted 1 unit to the right; if $c = 2$, the original graph is shifted 2 units to the left.

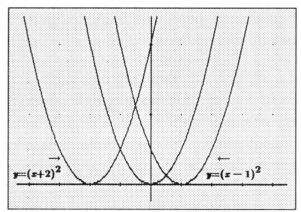

FIGURE 2.8. The graphs of $y = x^2$, $y = (x + 2)^2$, and $y = (x - 1)^2$.

What can you deduce about the effect of the value c? What is the relationship between the graphs of $y = f(x)$ and $y = f(x+c)$? See Exercise 2.

The graphs of $y = 0.5x^4$, $y = 0.5x^4 + 1$ and $y = 0.5x^4 - 2$ appear in Figure 2.9.

```
Range
Xmin:  -4.7
  max:   4.7
  scl:   1.
Ymin:  -3.1
  max:   3.1
  scl:   1.
```

The original function, $y = 0.5x^4$, crosses the y-axis at the point (0,0).

The function $y = 0.5x^4 + 1$ is the same shape but shifted up one unit.

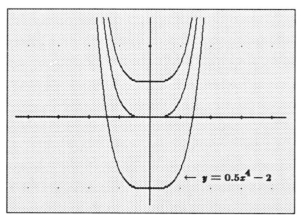

FIGURE 2.9. Vertical displacement of the function $f(x) = 0.5x^4$.

What is the effect of c in the graph of $y = f(x) + c$? See Exercise 3.

<u>Procedure 4.</u> We conclude this experiment with a look at the function $y = |x|$ and two variations. This is one graph that should be memorized because of its importance later.[7]

```
Range

Xmin:   -4.7
  max:   4.7
  scl:   1.
Ymin:   -1.5
  max:   6.
  scl:   1.
```

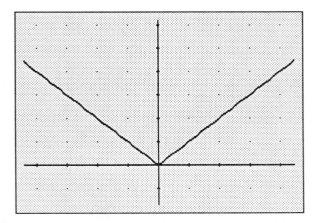

FIGURE 2.10. The absolute value function: $y = |x|$.

Two variations of the absolute value function appear below:

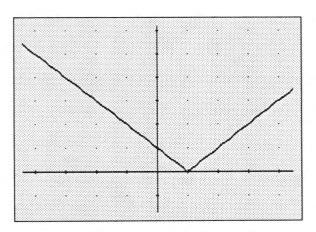

FIGURE 2.11. $y = |x - 1|$.

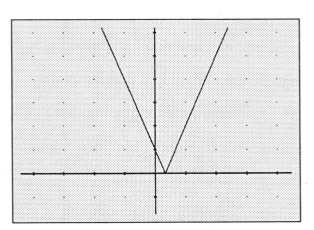

FIGURE 2.12. $y = |1 - 3x|$.

---

[7]The absolute value function is provided (It's the $\boxed{\text{SHIFT}}$ variation of the $\boxed{x^y}$ key.) Caution, however, you should be in COMP mode; the Abs function is not available in the SD1, SD2, LR1, or LR2 modes.

Notice that in the case of $y = |1 - 3x|$, for example, we have the line $y = 1 - 3x$ if y (or $1 - 3x$) is non-negative. That is $1 - 3x \geq 0$ (which, when solved for x means $x \leq \frac{1}{3}$). So, if $x \leq \frac{1}{3}$ the graph is actually the line $y = 1 - 3x$.

Since it is true that when $a < 0$, $|a| = -a$, then for the case $1 - 3x < 0$, we can write $|1 - 3x|$ as $-(1 - 3x) = 3x - 1$. Now $1 - 3x < 0$ means $x > \frac{1}{3}$ and we can say that for $x > \frac{1}{3}$, the graph is $y = 3x - 1$. In other words, if $x > \frac{1}{3}$, the graph in Figure 2.12 is the line $y = 3x - 1$.[8] Finally, the reader is urged to graph both $y = 3x - 1$ and $y = 1 - 3x$ and compare them to the graph in Figure 2.12.

---

[8] To graph $y = |1 - 3x|$ on the Casio, enter the following key sequence: Graph SHIFT Abs (1 − 3 ALPHA X) EXE.

Name _____   Experiment #2

Date _____   Exercise Sheet

DIRECTIONS: Show all work for each question in the space provided right on these sheets. For questions that require graphs, make the best sketch you can from the Casio's screen.

1. Consider the function $y = x^2 - 3x + 1$ in the interval $-1 \leq x \leq 2$. Graph the function on your calculator and determine the range for y for that interval. (See Procedure #1.)

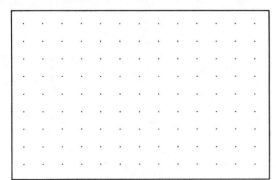

Range of y values: _____

2. In concise English sentences, describe the effect of the constant c on the graph of $y = f(x + c)$. (See Procedure #3 and the graphs in Figure 2.8.

_____

_____

3. In concise English sentences, describe the effect of the constant c on the graph of $y = f(x) + c$. (See Procedure #3 and the graphs in Figure 2.9.

_____

_____

In Exercises 4 through 6, graph each function in the right-hand column. Then match it with each of the graphs in Figures 2.4 through 2.7. Your answer should be a letter.

	Graphs:	*Functions:*
_____4.	Figure 2.4	a) $y = 2x^2 - 3$
_____5.	Figure 2.5	b) $y = 2x^2 - 3x$
_____6.	Figure 2.6	c) $y = x - x^3$
_____7.	Figure 2.7	d) $y = x^3 + 1$

In questions 8 through 10, determine whether the given function is even, odd, or neither by graphing it on your Casio graphics calculator. Circle the appropriate response.

8. $y = \dfrac{1}{x^2 - 4}$ ODD   EVEN   NEITHER

9. $y = \dfrac{1}{x^5}$ ODD   EVEN   NEITHER

10. $y = 7x^4 + 3x^2 + 5$ ODD   EVEN   NEITHER

11. Graph $y = x + \frac{1}{x}$ on your Casio calculator and use that to graph $y = x + \frac{1}{x} + 2$ below:

```
Range
Xmin:
 max:
 scl:
Ymin:
 max:
 scl:
```

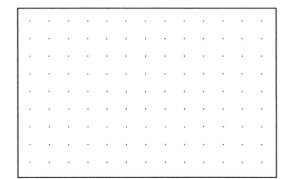

12. Graph $y = \dfrac{3x^2}{7}$ on your graphics calculator and use that to graph $y = \dfrac{3(x-3)^2}{7}$.

```
Range
Xmin:
 max:
 scl:
Ymin:
 max:
 scl:
```

13. Graph $y = |x^2|$ and $y = x^2$. In no more space than is provided below, write in your own words *why* there is no difference in these graphs.

_____

_____

14. a. An important graph in the study of calculus is $y = x^{2/3}$. Make a sketch of the graph of this function below. $\left[\text{HINT: Enter it as } \boxed{(} \ \boxed{\text{ALPHA}} \ \boxed{\text{X}} \ \boxed{x^2} \ \boxed{)} \ \boxed{x^y} \ \boxed{(} \ 1 \ \boxed{\div} \ 3 \ \boxed{)} \right]$

b. How does this graph compare with the graph of $\boxed{\text{ALPHA}} \ \boxed{\text{X}} \ \boxed{x^y} \ \boxed{(} \ 2 \ \boxed{\div} \ 3 \ \boxed{)}$? Why are the graphs different?

_____

_____

_____

Inverse Functions

INTRODUCTION

In this experiment, we will use the Casio graphics calculator as a means of determining whether or not two functions are inverses of each other. Recall, two functions are inverses of each other if the graph of one of the functions is the reflection of the other's graph about the line $y = x$.

PROCEDURES

Consider the function $f(x) = \dfrac{3 - 2x}{4}$. Your mathematics instructor has asked you to find the inverse of this function and you *think* your algebra is correct when you got $f^{-1}(x) = \dfrac{3 - 4x}{2}$. To check that result on the Casio graphics calculator, we graph each (linear) function and the line $y = x$ using the default range settings. If the functions are, in fact, inverses their graphs should be "mirror images" about the line $y = x$. Figure 3.1 depicts the three graphs.

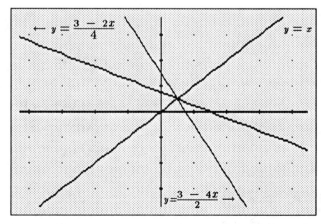

Yes, since the graph of $y = \dfrac{3 - 2x}{4}$ and the graph of $y = \dfrac{3 - 4x}{2}$ are, in fact, reflections about $y = x$, they are inverse functions.

FIGURE 3.1. $f(x) = \dfrac{3 - 2x}{4}$, its inverse, and $y = x$.

If the function is more complicated, the process of solving for the independent variable can be a difficult process. For example, consider $f(x) = \sqrt{2x - 3}$. It can be shown that the inverse is $f^{-1}(x) = \dfrac{x^2 + 3}{2}$ provided $x \geq 0$. On the Casio, we can force that restricted domain by setting the Xmin at 0 and the Xmax at some appropriate value, say 5.

Range	
Xmin:	0.0
max:	5.0
scl:	1.
Ymin:	0.0
max:	5.0
scl:	1.

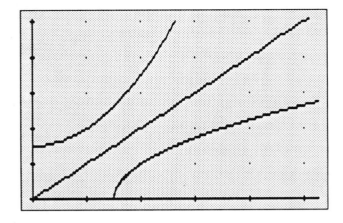

Again, we see that f^{-1} is the reflection of f about $y = x$.

FIGURE 3.2. $y = \sqrt{2x - 3}$, its inverse and $y = x$.

A Programming Consideration

The graph of a function can be thought of as a collection of points, (p, q). The inverse of the function (if it exists) is merely the collection of points (q, p). Enter the following program into your Casio to plot the function $y = \sqrt{2x - 3}$ and its inverse:

Program # P2

Logic of what must be done:	*Casio syntax*
1. Establish prompts for the user to enter the max and min settings for X and Y	1. "XMIN"?→ A: "XMAX" ?→ B: "YMIN" ?→ C: "YMAX" ?→D :
2. Assign an initial value to storage register P. (This is the initial, minimum X value and has already been stored in A.)	2. A → P
3. Calculate the Y value of the point and assign that value to register Q.	3. Lbl 1 : $\sqrt{2P - 3}$ → Q :
4. Plot the point (P, Q) -- a point on the function and also plot	4. PLOT P, Q :

Experiment #3
Page 28

(Q, P), a point on the inverse.   PLOT Q, P :

5. Increment the P value [9] and check to make sure it doesn't   5. P+(B − A) ÷ 94 → P :
exceed the maximum x value in storage register B; if it doesn't,   P < B ⇒ Goto 1 ◄
repeat steps 3, 4, and 5.

6. To display the graph at program's end, we add the "trick" of   6. Graph Y = 0
graphing $y = 0$.

    You should enter this program as program number P2. In the exercises, you will be
asked to alter it to accommodate other functions. For the program above, be careful upon
execution to make sure the X values are greater than 1.5 (WHY?). With Xmin at 1.5, Xmax
at 10, Ymin at 0 and Ymax at 5, we obtained the following graphs:

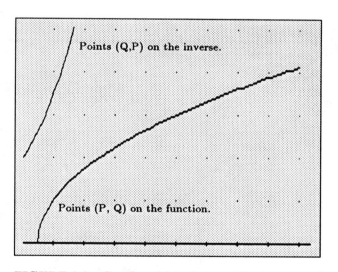

FIGURE 3.3. Graphs of function and its inverse using
program listing above.

---

[9]The Casio graphing screen is composed of 95 pixels from left to right. To increment
x values, we add the length of the X's (Xmax − Xmin) divided by 94 (there are 94 intervals
between pixels) to the previous x value.

Of course what makes the program especially useful is that all you need to do to is to change the first statement in Lbl 1 and the program will work for a new function. For example, if you wanted to graph $y = x^2$ and its inverse for $x \geq 0$, change "$\sqrt{2P - 3} \to Q$:" to "$P^2 \to Q$:" and let the program take care of the rest! The display of the output for range input of Xmin = 0, Xmax = 6, Ymin = 0, Ymax = 5 appears in Figure 3.4.

Range	
Xmin:	0.
max:	6.
scl:	1.
Ymin:	0.
max:	5.
scl:	1.

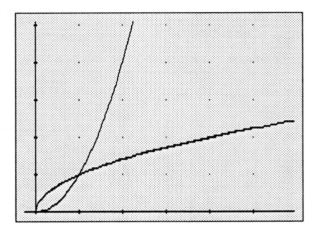

FIGURE 3.4. $y = x^2$ $(x \geq 0)$ and its inverse.

Name _____

Date _____

DIRECTIONS: Show all work to each question right on these pages. For graphs, make a sketch based on the Casio's graphing screen display.

In Exercises 1 through 6, find the inverse for each function. Then, check your work by graphing both the function and the inverse along with the line $y = x$ as in Figure 3.1.

1. $f(x) = 4x$

2. $f(x) = \sqrt{4 - x^2} \quad 0 \leq x \leq 2$

$f^{-1}(x) = $ _____

$f^{-1}(x) = $ _____

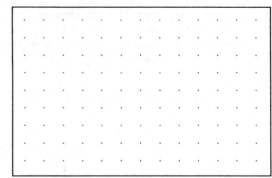

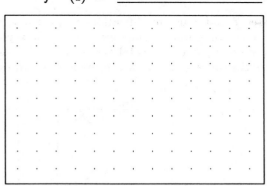

3. $f(x) = \dfrac{4}{x}$

4. $f(x) = \sqrt[3]{x - 1}$

$f^{-1}(x) = $ _____

$f^{-1}(x) = $ _____

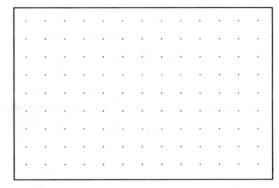

5. $f(x) = x^{3/5}$

$f^{-1}(x) =$ _____

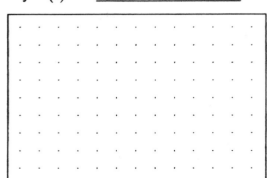

6. $f(x) = x^3 + 2$

$f^{-1}(x) =$ _____

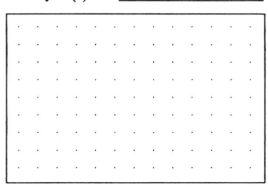

In Exercises 7 and 8 use modifications of program P2 from this section to graph the function and its inverse. Use appropriate range settings so that you can insure the function (with the appropriate restricted domain) does in fact have an inverse. Be sure to label both the function and its inverse. Also, state the range settings chosen upon program execution.

7. $f(x) = (x + 3)^2$

8. $f(x) = x^3 - x$

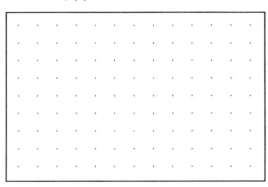

9. What would happen if you entered -1 as Xmin's value upon program execution for the function we first used in program P2? How could the program be modified to allow such input?

10. Consider $f(x) = \dfrac{1}{1 + x}$, $x \geq 0$. What is the domain of the inverse of f? [HINT: Use program P2 (modified for this function) and observe the inverse's domain.] Sketch the graph below.

Domain of f^{-1}: _____

Range of f: _____

Quadratic Functions

INTRODUCTION

1. A function of the form $y = Ax^2 + Bx + C$ is called a <u>quadratic function</u> whose graph is a <u>parabola.</u> If A > 0, the parabola opens upward; if A < 0, the parabola opens downward.

2. In the case of parabolas opening upward, the minimum value of the function (i.e., y value) is found at the parabola's <u>vertex</u>. If the parabola opens downward, it is the maximum value that occurs at the vertex. The coordinates of the vertex are $\left(\frac{-B}{2A}, \ -\frac{B^2 - 4AC}{4A} \right)$.

3. The <u>axis of symmetry</u> of a parabola is the vertical line through the vertex. The parabola is symmetric with respect to that line.

PROCEDURES

<u>Procedure 1.</u> By now, it is an easy task for us to graph just about any function on the Casio. In particular, to consider the quadratic function $y = 2x^2 - 3x + 7$ we merely have to enter the following keystrokes:

$\boxed{\text{Graph}}$ 2 $\boxed{\text{ALPHA}}$ $\boxed{\text{X}}$ $\boxed{x^2}$ $\boxed{\ -\ }$ 3 $\boxed{\text{ALPHA}}$ $\boxed{\text{X}}$ $\boxed{+}$ 7 $\boxed{\text{EXE}}$.

But nothing appears assuming the default range settings are in effect! This can take some experimentation. For now, try entering the following settings: Xmin: -4.7, Xmax: 4.7, Xscl: 1, Ymin: 2, Ymax: 12, Yscl: 2. The following graph should appear:

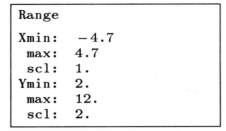

```
Range
Xmin:   −4.7
 max:    4.7
 scl:    1.
Ymin:    2.
 max:    12.
 scl:    2.
```

For clarity, we have marked the y axis with some references. It is important to keep in mind max and min values when observing a graph.

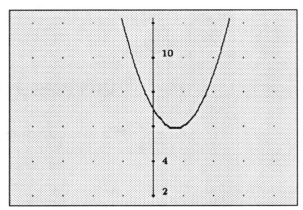

FIGURE 4.1. $y = 2x^2 - 3x + 7.$

Procedure 2. To generalize the graphing of a quadratic function, we will use program P3 below. We will use it to graph the function $y = -2x^2 + 3x - 7$ (the negative of the previous quadratic function). Notice, as in program P2, that we let the range settings be entered by the user during program execution.

$$\boxed{\text{Program \# P3}}\ ^{10}$$

Logic of what must be done: *Casio syntax*

1. Establish prompts for the user to enter the max and min values for X and Y. We use registers E, F, G, and H to avoid confusion with A, B, and C (the function's coefficients).

1. "XMIN"? → E :
    "XMAX" ? → F :
    "YMIN" ? → G :
    "YMAX" ? → H :

2. Here, we allow for a horizontal and vertical scaling factor.

2. "X SCL" ? → I :
    "Y SCL" ? → J :

3. Enter the range. (Be careful with the order of the storage registers!)

3. Range E, F, I, G, H, J

4. Enter the coefficients A, B, and C and store them in registers with the same name.

4. "A = " ? → A :
    "B = " ? → B :
    "C = " ? → C :

5. Graph the function.

5. Graph Y $= AX^2 + BX + C$ ◢

6. Calculate the coordinates of the vertex and display them.

6. "VERTEX" ◢
    $-B \div (2A) \to X$ ◢
    $AX^2 + BX + C$ ◢

---

[10]Note: You may have to clear old programs if there is not enough space for program P3. This is especially appropriate on the *fx-7000G* which only has 422 bytes of memory. The *fx-7500G* contains over 4000 bytes of memory and can hold all the programs in this manual without a problem.

As an example of program P3's execution, we entered the range settings as they appear in Figure 4.2. Also, in order to complete the program's execution, after the graph is displayed, press $\boxed{\text{G}\leftrightarrow\text{T}}$ to see the text screen again and then hit $\boxed{\text{EXE}}$ three times to see the coordinates of the vertex; follow that display by the $\boxed{\text{AC}}$ key to clear the text screen. The output for a run of program P3 appears in Figures 4.2 and 4.3.

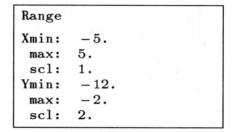

Compare this graph with Figure 4.1. In general, the graph of $y = -f(x)$ is the reflection of the graph of $y = f(x)$ about the x-axis.

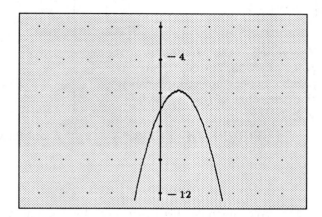

FIGURE 4.2. The graph of $y = -2x^2 + 3x - 7$ and its associated range settings.

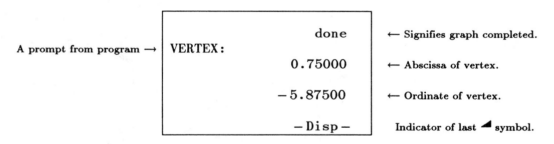

FIGURE 4.3. The text screen at the end of running program P3.

Procedure 3. The axis of symmetry for the parabola in Figure 4.2 is $x = 0.75$ because the axis of symmetry has to pass through the vertex. To graph it along with the parabola, we can modify program P3 to include the vertical line plotting capability from program P1. Add the following lines to the end of program P3:

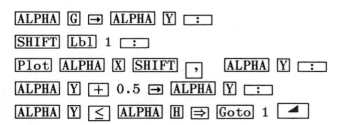

This new version of program P3 was used to plot $y = 3x^2 - 5x + 6$ along with its axis of symmetry. The result is displayed in Figure 4.4.

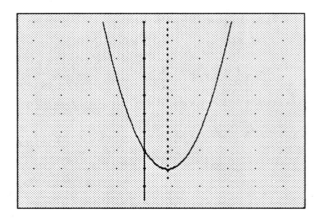

```
Range
Xmin:   -3.
  max:   5.
  scl:   1.
Ymin:   3.
  max:  20.
  scl:   2.
```

The vertex of the parabola is at the point (0.83333, 3.91667). Since the parabola opens upward, we can say that the range of this function is $y \geq 3.91667$.

FIGURE 4.4. $y = 3x^2 - 5x + 6$ and its axis of symmetry, $x = 0.83333$.

We conclude this experiment by reminding the reader that an alternative to finding the vertex would be to use the $\boxed{\text{Trace}}$ key; that method is less precise, however, due to the discrete nature of the pixels on the graphing screen.

Name _____ Experiment #4

Date _____ Exercise Sheet

DIRECTIONS: Show all work to each question right on these pages. For graphs, make a sketch based on the Casio's graphing screen display.

In questions 1 through 5, plot the quadratic function and its axis of symmetry. State the range settings you had the Casio use for this experiment. Also, state the vertex and the axis of symmetry.

1. $y = x^2 + 8x + 11$

```
Range
Xmin:
 max:
 scl:
Ymin:
 max:
 scl:
```

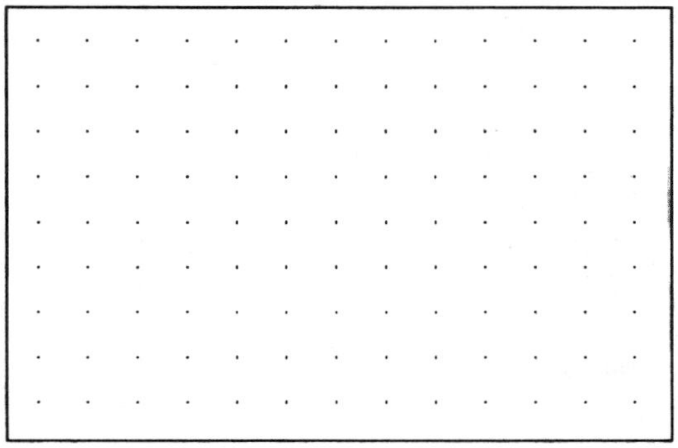

Coordinates of vertex: _____   Axis of symmetry: _____

2. $y = -x^2 - 4x + 1$

```
Range
Xmin:
 max:
 scl:
Ymin:
 max:
 scl:
```

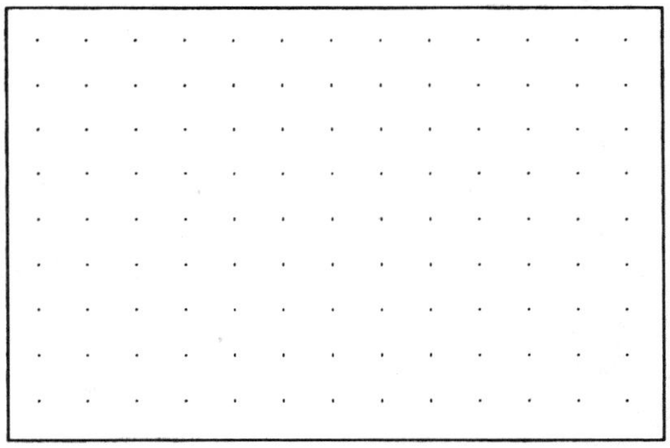

Coordinates of vertex: _____   Axis of symmetry: _____

3. $y = 4x^2 - 4x + 21$

```
Range
Xmin:
 max:
 scl:
Ymin:
 max:
 scl:
```

Coordinates of vertex: _____   Axis of symmetry: _____

4. $y = 25 - x^2$

```
Range
Xmin:
 max:
 scl:
Ymin:
 max:
 scl:
```

Coordinates of vertex: _____   Axis of symmetry: _____

5. $y = x^2 + 3x + \frac{1}{4}$

```
Range
Xmin:
 max:
 scl:
Ymin:
 max:
 scl:
```

Coordinates of vertex: _____   Axis of symmetry: _____

Polynomial Functions of Higher Degree

INTRODUCTION

1. The general n^{th} degree polynomial function can be written as

$$y = a_n x^n + a_{n-1} x^{n-1} + \cdots + a_1 x + a_0.$$

The number a_n is called the <u>leading coefficient</u> of the polynomial and n determines the <u>degree</u> of the polynomial. If the degree is odd, the graph of the function must cross the x-axis at least once. Equivalently, we say that the polynomial $a_n x^n + a_{n-1} x^{n-1} + \cdots + a_1 x + a_0$ has at least one real *zero*.

2. We can use the graphics calculator to *approximate* zeros of a polynomial by examining where the graph crosses the x-axis. Then, using zooming techniques, we can refine that approximation.

3. The graphics calculator can also help us observe where a function (the y-value) is positive and where it is negative.

PROCEDURES

<u>Procedure 1.</u> In general, as the degree of a polynomial function increases, its graph tends to exhibit more peaks and valleys (or *turning points*). Also, if the leading coefficient is positive, the graph will rise up to the right; if the leading coefficient is negative, the graph will fall to the right. To see this, we have graphed some polynomial functions in the next eight figures.

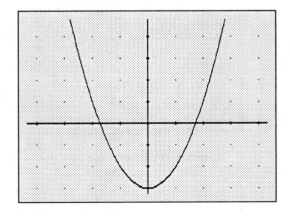

FIGURE 5.1. Second degree polynomial with $a_n > 0$.

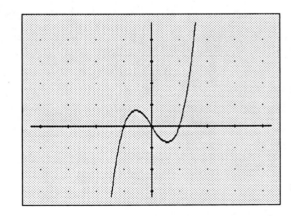

FIGURE 5.2. Third degree polynomial with $a_n > 0$.

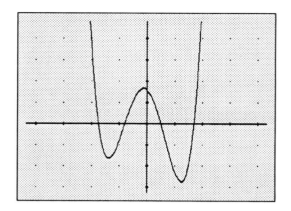

FIGURE 5.3. Fourth degree polynomial with $a_n > 0$.

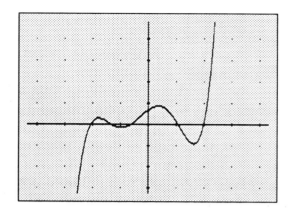

FIGURE 5.4. Fifth degree polynomial with $a_n > 0$.

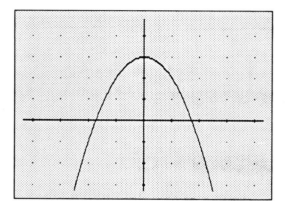

FIGURE 5.5. Second degree polynomial with $a_n < 0$.

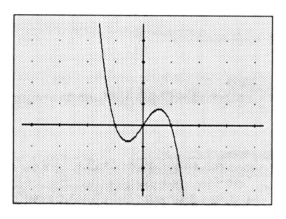

FIGURE 5.6. Third degree polynomial with $a_n < 0$.

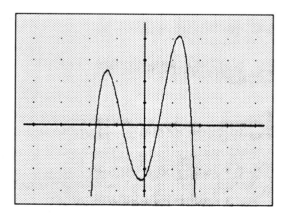

FIGURE 5.7. Fourth degree polynomial with $a_n < 0$.

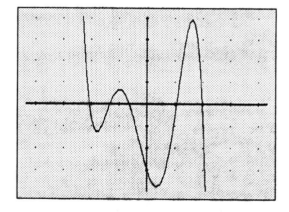

FIGURE 5.8. Fifth degree polynomial with $a_n < 0$.

<u>Procedure 2.</u> Let's use the Casio graphics calculator to graph $y = x^3 + x^2 - 3x + 3$ and attempt to approximate any real zeros of $x^3 + x^2 - 3x + 3$. (That is, we are looking for values(s) of x that will make the polynomial zero. These are precisely the places where the graph of the polynomial function crosses the x-axis.) The range and graph appear in Figure 5.9.

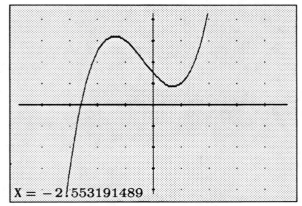

```
Range
Xmin:   -5.
 max:    5.
 scl:    1.
Ymin:   -8.
 max:    8.
 scl:    2.
```

There appears to be one x-intercept for this function. Using the $\boxed{\text{Trace}}$ key, we see that it is approximately -2.55.

FIGURE 5.9. The graph of $y = x^3 + x^2 - 3x + 3$.

To see how we can refine this approximation, we will use the Casio's ZOOM-IN capability. With the $\boxed{\text{Trace}}$ key depressed and using the cursor movement keys, move to the x value closest to the x axis on the curve. If you used our range settings above, that should be -2.553191489. Now press $\boxed{\text{SHIFT}}$ $\boxed{\text{X}}$. Notice that the calculator automatically zooms in using the traced point as the center of the graphing screen. Do it again. This twice zoomed-in variation is graphed in Figure 5.10.

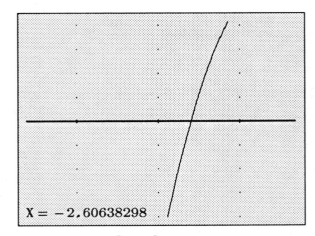

```
Range
Xmin:   -3.85638298
 max:   -1.35638298
 scl:    1.
Ymin:   -2.12903226
 max:    1.87096774
 scl:    2.
```

The x value where the curve appears to be closest to the x-axis is between -2.60638298 and -2.579787235.

FIGURE 5.10. Zoomed-in view of $y = x^3 + x^2 - 3x + 3$.

By more advanced (algebraic) techniques, we can find that the *exact* value of the zero is

$\frac{1}{3}(\sqrt[3]{10} - \sqrt[3]{100} - 1) \approx -2.59867451$. We'll approximate that zero as -2.6.

Procedure 3. Look carefully at the graph in Figure 5.9. Points on the graph that are above the x-axis are points where y is positive. Similarly, where the curve falls below the x-axis, y values are negative.

In a fashion similar to the zooming technique used above to find x-intercept(s), we can also find turning points of the curve. These points determine the intervals on which the function is increasing (graphically, movement from lower left toward upper right) and where it is decreasing (movement from upper left to lower right). The reader is urged to verify the points highlighted in Figure 5.11.

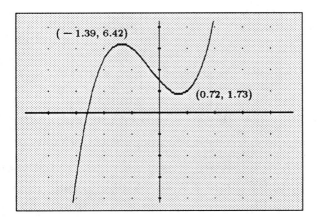

The two turning points have coordinates $(-1.39, 6.42)$ and $(0.72, 1.73)$. (These have been rounded off and were obtained by zooming in several times on the points.

The domain and range of this cubic polynomial are both the collection of all real numbers, **R**.

FIGURE 5.11. $y = x^3 + x^2 - 3x + 3$ has one "peak" and one "valley".

We summarize the function's behavior below:

If $x < -2.6$, the function (or y-values) is negative.

If $x > -2.6$, the function is positive.

If $x < -1.39$, the function is increasing.

If $-1.39 < x < 0.72$, the function is decreasing.

If $x > 0.72$, the function is increasing.

Name _____   **Experiment #5**

Date _____   **Exercise Sheet**

DIRECTIONS: Show all work to each question right on these pages. For graphs, make a sketch based on the Casio's graphing screen display.

In questions 1 through 4, graph the given polynomial function. Also, using the Trace key and the zooming feature of the Casio graphics calculator, approximate zeros of the polynomial and coordinates of any turning points. State where (x intervals) the function is positive, negative, increasing, and decreasing.

1. $y = 3x^4 + 4x^3$

    Zeros: _____

    Turning points:

    _____

    _____

    _____

For what values of x is the function positive? _____

For what values of x is the function negative? _____

For what values of x is the function increasing? _____

For what values of x is the function decreasing? _____

2. $y = 3x^3 - 9x + 1$

    Zeros: _____

    Turning points:

    _____

    _____

    _____

For what values of x is the function positive? _____

Notice that the function has a value of zero when $x = \frac{1}{2}$ [the x-intercept is $(\frac{1}{2}, 0)$]; that is the value of x for which the numerator is zero. The vertical asymptote occurs at $x = 0$, the x-value that makes the denominator zero.

The next example we would like to graph is $y = \dfrac{2x}{x^2 - 1}$. The same range settings are in effect.

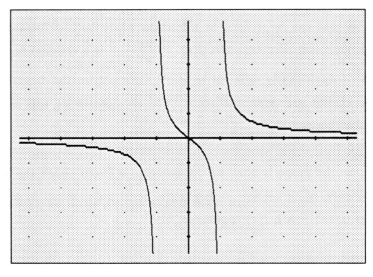

There are two vertical asymptotes: $x = -1$ and $x = 1$.
The y-axis (or $x = 0$) is the only horizontal asymptote.
The domain of this function is $x \neq \pm 1$; its range is **R**.

FIGURE 6.2. $\mathbf{y} = \dfrac{2x}{x^2 - 1}$.

It should be noted that in the previous example, the curve crosses its asymptote. Also, $f(x) = \dfrac{2x}{x^2 - 1}$ is an odd function (the curve is symmetric with respect to the origin) and its only intercept is $(0, 0)$.

<u>Procedure 2.</u> Next, we will graph $y = \dfrac{2(x^2 - 9)}{x^2 - 4}$. Note the range settings.

Range	
Xmin:	$-8.$
max:	$8.$
scl:	$2.$
Ymin:	$-12.$
max:	$12.$
scl:	$3.$

Domain: $x \neq \pm 2$
Range: $y < 2$ or $y \geq \frac{9}{2}$
Horizontal asymptote: $y = 2$
Vertical asymptotes: $x = -2$, $x = 2$

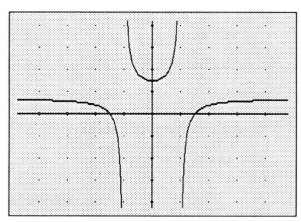

FIGURE 6.3. $\mathbf{y} = \dfrac{2(x^2 - 9)}{x^2 - 4}$.

The function $f(x) = \dfrac{2(x^2 - 9)}{x^2 - 4}$ is even (its graph is symmetric with respect to the y-axis); its intercepts are $(0, \frac{9}{2})$, $(-3, 0)$, and $(3, 0)$.

We conclude this experiment by graphing the (odd) rational function $y = \dfrac{x}{x^2 + 1}$. We graph it in the next two figures with different Casio range settings to emphasize the importance of that feature.

```
Range
Xmin:   -5.
 max:    5.
 scl:    1.
Ymin:   -4.
 max:    4.
 scl:    2.
```

There is no need for y values to vary between -4 and 4. The graph would be easier to read if we were to "spread" the y-axis out.

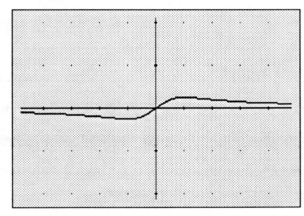

FIGURE 6.4. $y = \dfrac{x}{x^2 + 1}$.

```
Range
Xmin:   -5.
 max:    5.
 scl:    1.
Ymin:   -1.
 max:    1.
 scl:    0.5
```

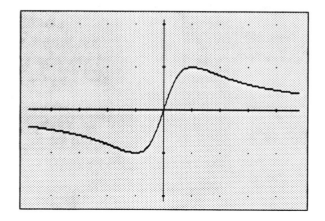

Domain: **R**
Range: $-0.5 \leq y \leq 0.5$
The only intercept is $(0, 0)$.
Horizontal asymptote: $y = 0$

FIGURE 6.5. $y = \dfrac{x}{x^2 + 1}$.

The function $f(x) = \dfrac{x}{x^2 + 1}$ has no vertical asymptotes. The reason is this: there are no values of x which make the denominator of the rational function close to zero (and hence the value of the fraction infinitely large). Remember, as you divide by numbers closer and closer to zero, the value of the fraction (with non-zero numerator) gets infinitely large.

Name _____   Experiment #6

Date _____   Exercise Sheet

DIRECTIONS: Show all work to each question right on these pages. For graphs, make a sketch based on the Casio's graphing screen display.

In questions 1 through 5, use the Casio graphics calculator to make a sketch of the given rational function. Also, state its domain, range, vertical asymptotes, horizontal asymptotes, intercepts, and symmetry (if any).

1. $y = \dfrac{x + 1}{x - 2}$

        Domain: _____

        Range: _____

        Vertical Asymptotes: _____

        Horizontal Asymptotes: _____

        Intercepts: _____

        Symmetry: _____

2. $y = \dfrac{x + 3}{(x - 2)(x + 4)}$

        Domain: _____

        Range: _____

        Vertical Asymptotes: _____

        Horizontal Asymptotes: _____

        Intercepts: _____

        Symmetry: _____

3. $y = \dfrac{x}{x^2 - 2}$

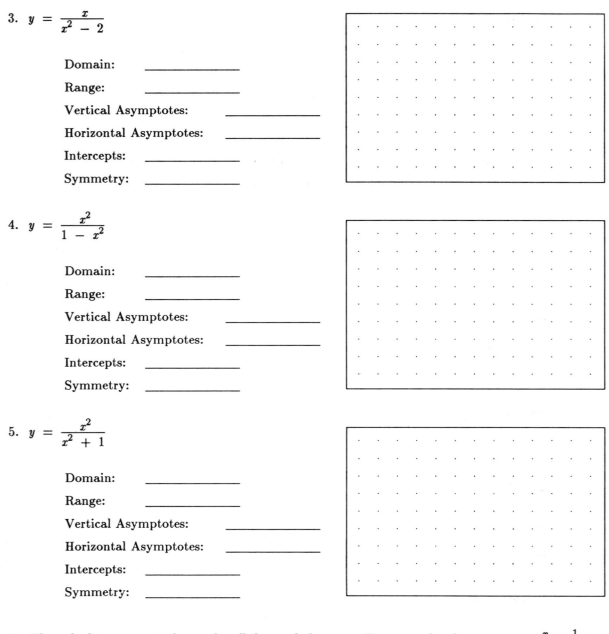

      Domain: _____

      Range: _____

      Vertical Asymptotes: _____

      Horizontal Asymptotes: _____

      Intercepts: _____

      Symmetry: _____

4. $y = \dfrac{x^2}{1 - x^2}$

      Domain: _____

      Range: _____

      Vertical Asymptotes: _____

      Horizontal Asymptotes: _____

      Intercepts: _____

      Symmetry: _____

5. $y = \dfrac{x^2}{x^2 + 1}$

      Domain: _____

      Range: _____

      Vertical Asymptotes: _____

      Horizontal Asymptotes: _____

      Intercepts: _____

      Symmetry: _____

6. The calculator cannot always do *all* the work for you. For example, the curve $y = \dfrac{x - 1}{x^2 - 1}$ is not defined for $x = \pm 1$. However, when the Casio plots this function, it may or may not plot the point $(1, \frac{1}{2})$, depending on your range settings. Technically, there should be a *puncture hole* at $(1, \frac{1}{2})$. Why? Where does the value $\frac{1}{2}$ come from?

_____

_____

Exponential Functions

INTRODUCTION

1. An <u>exponential function</u> with base b is $f(x) = b^x$ where $b > 0$ and $b \neq 1$. If $b > 1$, the function is said to exhibit <u>exponential growth</u>; if $0 < b < 1$, the function exhibits <u>exponential decay.</u>

2. The <u>natural exponential function</u> has base $e \approx 2.718281828$.

3. More complicated exponential functions, such as $y = 2^{1-x^2}$, can be graphed with ease using the Casio graphics calculator. Choosing appropriate range settings is the only tricky part.

4. A very common application of exponential functions is <u>compound interest</u>. The amount (or value), A, of an investment of P dollars at annual interest rate r compounded n times per year for t years is given by:

$$A = P\left(1 + \tfrac{r}{n}\right)^{nt}$$

If the compounding is *continuous,* the formula becomes: $A = Pe^{rt}$.

PROCEDURES

<u>Procedure 1.</u> We will use the Casio to graph $y = 3^x$. It is depicted in Figure 7.1. Notice that it exhibits exponential growth, its intercept is $(0, 1)$, and its range is $y > 0$. This last fact causes us to use Casio graphing range settings that start just below the x-axis.

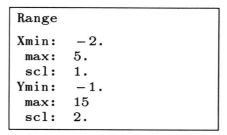

Range

Xmin: −2.
 max: 5.
 scl: 1.
Ymin: −1.
 max: 15
 scl: 2.

Exponential growth.

Domain: **R**

Range: $y > 0$

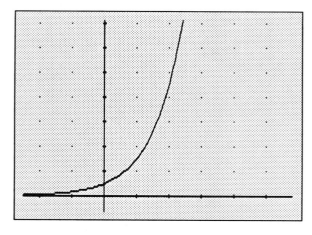

FIGURE 7.1. $y = 3^x$

Procedure 2. Now, we will graph a curve exhibiting exponential decay, $y = \left(\frac{1}{2}\right)^x$. Keep in mind, this is exactly the same function as $y = 2^{-x}$.

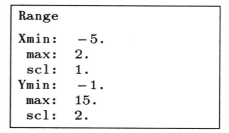

Range

Xmin: −5.
 max: 2.
 scl: 1.
Ymin: −1.
 max: 15.
 scl: 2.

Exponential decay

Domain: **R**

Range: $y > 0$

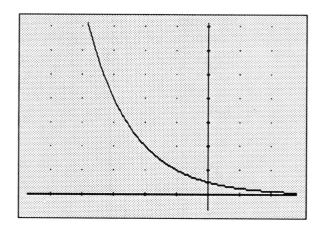

FIGURE 7.2. $y = \left(\frac{1}{2}\right)^x$

Procedure 3. A more complicated exponential function is $y = 2^{(1-x^2)}$. The sequence for graphing it, once the appropriate range settings have been entered, are:

$$\boxed{\text{Graph}}\ 2\ \boxed{x^y}\ \boxed{(}\ 1\ \boxed{-}\ \boxed{\text{ALPHA}}\ \boxed{\text{X}}\ \boxed{x^2}\ \boxed{)}\ \boxed{\text{EXE}}$$

The graph is displayed in Figure 7.3.

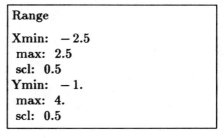

Range

Xmin: -2.5
 max: 2.5
 scl: 0.5
Ymin: $-1.$
 max: $4.$
 scl: 0.5

Even function (y-axis symmetry)

Domain: **R**

Range: $0 < y \leq 2$

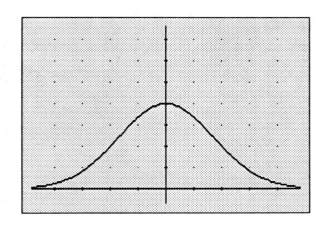

FIGURE 7.3. $y = 2^{(1-x^2)}$

Procedure 4. The number e is the most commonly used base in calculus. The functions $y = e^x$ and $y = e^{-x}$ appear in Figure 7.4.

Range

Xmin: $-3.$
 max: $3.$
 scl: $1.$
Ymin: $-1.$
 max: $6.$
 scl: $1.$

Notice both curves pass through the point (0, 1).

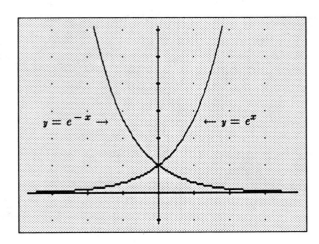

$y = e^{-x} \rightarrow$ $\leftarrow y = e^x$

FIGURE 7.4. Graphs of the functions $y = e^{-x}$ and $y = e^x$.

To have the greatest number of decimal digits displayed, press ⟦MODE⟧ 7 followed by 9 ⟦EXE⟧. Now, to evaluate powers of e, we will use the ⟦e^x⟧ key. For e^1, for example, press ⟦e^x⟧ 1. You should see 2.718281828 displayed. To evaluate e^2, press ⟦e^x⟧ 2 and 7.389056099 appears on the text screen.

<u>Procedure 5.</u>   We conclude this experiment with some practice with computations using the compound interest formula. Here is the problem:

> Marnie has \$9000 to invest in a savings account and she has narrowed her choices to two banks: Megabux Savings offers an annual percentage rate of 8.5% compounded quarterly. Frugal Federal offers an annual percentage rate of 8% compounded continuously. After two years, what will the \$9000 amount to in each bank?

Solution: Since we are dealing with monetary units, it suffices to display two decimal places. We can achieve this by pressing $\boxed{\text{MODE}}$ 7 2 $\boxed{\text{EXE}}$. We calculate each bank's interest as follows:

<u>Megabux Savings:</u> We use the compound interest formula $A = P\left(1 + \frac{r}{n}\right)^{nt}$ with $P = \$9000$, $r = 0.085$, $n = 4$, and $t = 2$. To find A on the Casio, key in the following:

$$9000 \;\boxed{\times}\; \boxed{(}\; 1 \;\boxed{+}\; 0.085 \;\boxed{\div}\; 4 \;\boxed{)}\; \boxed{x^y}\; \boxed{(}\; 4 \;\boxed{\times}\; 2 \;\boxed{)}\; \boxed{\text{EXE}}$$

At the end of two years, the Casio reports that the initial investment will be worth \$10,648.76.

<u>Frugal Federal:</u> Here we use the formula $A = Pe^{rt}$. So we have:

$$9000 \;\boxed{e^x}\; \boxed{(}\; 0.080 \;\boxed{\times}\; 2 \;\boxed{)}\; \boxed{\text{EXE}}$$

The initial investment is worth \$10,561.60 after two years of continuous compounding.

It appears that Marnie would maximize her investment if she went with Megabux Savings.

Name _____   Experiment #7

Date _____   Exercise Sheet

DIRECTIONS: Show all work to each question right on these pages. For graphs, make a sketch based on the Casio's graphing screen display.

In questions 1 through 6, sketch the graph of the given function. Also, state its domain and range.

1. $y = 5^x$ Domain: _____

    Range: _____

2. $y = (0.6)^x$ Domain: _____

    Range: _____

3. $y = 2^{-x^2}$ Domain: _____

    Range: _____

4. $y = \left(\frac{3}{2}\right)^{x+2}$ Domain: _____

    Range: _____

5. $y = 5^{|x|}$ Domain: _____

                   Range: _____

6. $y = e^{x-1}$ Domain: _____

                  Range: _____

7. We have discussed in a previous experiment that the graph of $y = f(x)$ is shifted c units vertically to obtain the graph of $y = f(x) + c$. Also, the graph of $y = f(x+c)$ represents a horizontal shift. Using that information, sketch a graph of $y = \left(\frac{1}{2}\right)^x + 3$ and $y = \left(\frac{1}{2}\right)^{x-2}$ based on the graph in Figure 7.2.

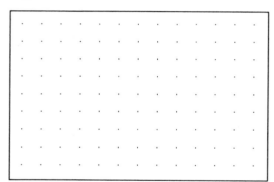 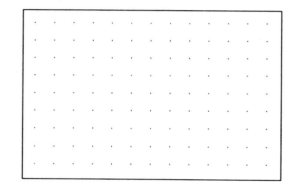

In questions 8 through 10, find the value of a $10,000 investment after t years at the given compounding frequency and annual percentage rates.

8. a. Rate = 9% $t = 2$ years with daily ($n = 365$) compounding   b. Continuous compounding

9. a. Rate = 7.5% $t = 3$ years with monthly compounding   b. Continuous compounding

10. a. Rate = 8.75% $t = 1$ year with quarterly compounding   b. Continuous compounding

Logarithmic Functions

INTRODUCTION

1. The <u>logarithmic function</u> is the inverse of the exponential function. We write $y = \log_b x$ as an equivalent expression for $x = b^y$ but solved for y. (We will assume $b > 1$.)

2. The number 10, is the base of the so-called "common" logarithm and we write $\log x$. The number e is the base of the "natural" logarithm and written $\ln x$. The numbers 10 and e are the two most often used bases. These correspond to the Casio's keys $\boxed{\log}$ and $\boxed{\ln}$. For bases other than these, we can use the <u>change of base</u> rule for logarithms:

$$\log_b x = \frac{\ln x}{\ln b} = \frac{\log x}{\log b}$$

PROCEDURES

<u>Procedure 1.</u> We graph the common logarithm, $y = \log x$ in Figure 8.1. Note the Casio range settings and the domain and range. (The domain and range correspond, of course, to the range and domain of the function $y = 10^x$.)

Range

Xmin: $-1.$
  max: 25.
  scl: 5.
Ymin: -2.
  max: 2.
  scl: 0.5

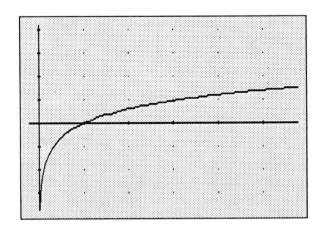

Domain: $x > 0$
Range: $\mathbb{R}$
Intercept: $(1, 0)$
Vertical asymptote: $x = 0$

FIGURE 8.1. $\mathbf{y} = \mathbf{\log_{10} x}.$

As with some of our other discussion regarding graphing, keep in mind that the logarithm function graph in Figure 8.1 would be shifted two units to the right to obtain the graph of $y = \log(x - 2)$ or three units vertically upward to obtain $y = 3 + \log x$.

Procedure 2. In Figure 8.2, we see the inverse relationship between the logarithmic and exponential functions. One is the mirror image of the other about the line $y = x$.

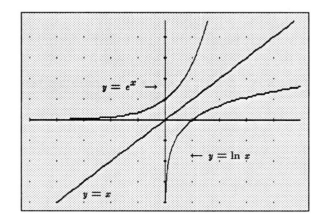

Range	
Xmin:	$-5.$
max:	$5.$
scl:	$1.$
Ymin:	$-3.$
max:	4
scl:	$1.$

	$y = e^x$	$y = \ln x$
Domain:	$\mathbb{R}$	$x > 0$
Range:	$y > 0$	$\mathbb{R}$
Intercepts:	$(0, 1)$	$(1, 0)$

FIGURE 8.2. The inverse functions $y = e^x$ and $y = \ln x$.

Calculations of logarithms are relatively straight forward on the Casio. Remember to use the change of base formula for bases other than 10 or e. For example, to calculate $\log_7 49$, enter the following keystrokes:

$$\boxed{\log}\ 4\ 9\ \boxed{\div}\ \boxed{\log}\ 7\ \boxed{\text{EXE}}$$

or

$$\boxed{\ln}\ 4\ 9\ \boxed{\div}\ \boxed{\ln}\ 7\ \boxed{\text{EXE}}$$

In either case, the result of 2.00000 is displayed.

Name _____   **Experiment #8**

Date _____   **Exercise Sheet**

DIRECTIONS: Show all work to each question right on these pages. For graphs, make a sketch based on the Casio's graphing screen display.

In questions 1 through 4, graph the given logarithmic function and state its domain and range.

1. $y = \log (x + 1)$

     Domain: _____

     Range: _____

2. $y = 4 + \ln x$

     Domain: _____

     Range: _____

3. $y = - \log x$

     Domain: _____

     Range: _____

4. $y = -4 + \ln x$

     Domain: _____

     Range: _____

5. The absolute value function is an interesting one to compose with other functions. Can you determine what the graph of $y = \ln |x|$ will look like? If necessary, use the Casio to graph it. How does the domain of this function differ from the natural logarithmic function?

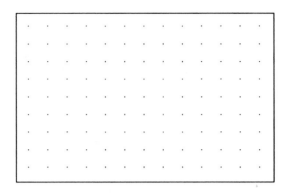

6. What does the graph of $y = |\ln x|$ look like? What are its domain and range?

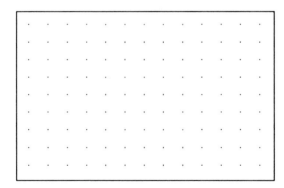

7. Recall the compound interest formula: $A = P\left(1 + \frac{r}{n}\right)^{nt}$. Use it to find the length of time (in years) it will take an investment to triple if the annual percentage rate is 7.75% and the interest is compounded monthly. [HINT: $A = 3P$. To solve for t means using logarithms.]

Approximating Solutions to Exponential and Logarithmic Equations

INTRODUCTION

1. Although it is a relatively easy task to solve certain contrived exponential or logarithmic equations, such as $2^x = 16$ or $\log(x-2) = 3$, more complicated equations are solved using approximation techniques. The Casio graphics calculator has a built-in "approximator" with its graphics screen. To solve an equation such as $e^{2x} = 1 - x^2$, we can look to see where the graphs of $y = e^{2x}$ and $y = 1 - x^2$ intersect.

2. Alternatively, we could graph the function $y = e^{2x} + x^2 - 1$ and determine where its x-intercepts are. Equivalently, that means we are solving $e^{2x} + x^2 - 1 = 0$. In either case, using the zoom-in technique discussed earlier, we can refine the approximation significantly.

PROCEDURES

<u>Procedure 1.</u> We want to approximate the solutions to the equation $e^{2x} = 1 - x^2$, hence we will graph $y = e^{2x}$ and $y = 1 - x^2$. First, we establish a short program to graph the two functions. With the calculator in program writing mode ($\boxed{\text{MODE}}$ 2) choose program #4. It is a simple two line program:

Enter for line 1: $\boxed{\text{Graph}}\,\boxed{e^x}\,2\,\boxed{\text{ALPHA}}\,\boxed{\text{X}}\,\boxed{\text{EXE}}$ Displayed syntax: Graph Y = e2X

Enter for Line 2: $\boxed{\text{Graph}}\,1\,\boxed{-}\,\boxed{\text{ALPHA}}\,\boxed{\text{X}}\,\boxed{x^2}$ Displayed syntax: Graph Y = 1 − X²

It appears that the two functions have two points in common and one of them, (0, 1), can be easily read from the graph. We are concerned, however, with estimating the coordinates of the other point of intersection and on the screen in Figure 9.1, it is very difficult to approximate it.

Range	
Xmin:	-4.7
max:	4.7
scl:	$1.$
Ymin:	-3.1
max:	3.1
scl:	$1.$

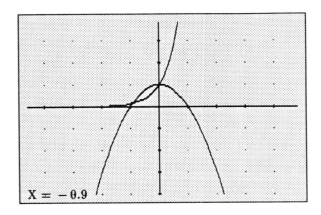

There are two points of intersection: $(0, 1)$ and, using the [Trace] key, somewhere around $x = -0.9$

FIGURE 9.1. To approximate solutions to $e^{2x} = 1 - x^2$, we graph $y = e^{2x}$ and $y = 1 - x^2$.

With the flashing pixel at $X = -0.9$ (from the trace command), press [SHIFT] [X] to zoom in. Repeat the process two more times, with the flashing pixel at $X = -0.95$ and then at $X = -0.925$. Notice that each time *both* the graphs get redrawn. (Now you understand the reason for using the program. Otherwise, just one function's graph would have been redrawn -- the one entered last.) The result of zooming in thrice is depicted in Figure 9.2.

Range	
Xmin:	-2.125
max:	0.225
scl:	$1.$
Ymin:	-0.675
max:	0.875
scl:	$1.$

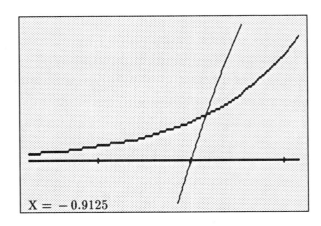

It appears that the approximation is $x \approx -0.9125$. Note that $e^{2x} \approx 0.1612$ and $1 - x^2 \approx 0.1673$.

FIGURE 9.2. The approximate solution to $e^{2x} = 1 - x^2$, is -0.9125.[12]

---

[12]It can be shown using advanced techniques that the x value -0.916562583 is a more precise approximation. This value assures a solution to within 10^{-9}.

It should be noted that all five of the following statements are equivalent and the reader should be familiar enough with the vocabulary of mathematics to understand these equivalences:

"-0.9125 is an approximate solution to the equation $e^{2x} = 1 - x^2$"

"-0.9125 is an approximate zero of $e^{2x} - 1 + x^2$"

"If $f(x) = e^{2x} - 1 + x^2$, then $f(-0.9125) \approx 0$"

"The graphs of $y = e^{2x}$ and $y = 1 - x^2$, intersect at a point whose abscissa is about -0.9125."

"The x-intercept of the graph of $y = e^{2x} - 1 + x^2$ is approximately $(-0.9125, 0)$."

This last interpretation is illustrated in Figures 9.3a and 9.3b below.

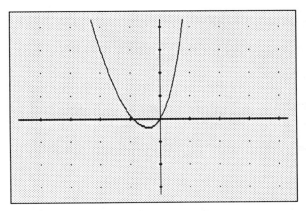

FIGURE 9.3a. $y = e^{2x} - 1 + x^2$ (default settings)   FIGURE 9.3b. After zooming in three times.

Procedure 2. We conclude this experiment by mentioning that the approximation technique discussed here can be applied to any function. Consider the problem of solving the logarithmic equation $\ln(2x - 1) = 3 - \frac{x}{2}$. We will approximate the solution(s) by graphing $y = \ln(2x - 1)$ and $y = 3 - \frac{x}{2}$ on the same set of coordinate axes and zooming in on the point of intersection. Figure 9.4. shows that 2.875 is the approximate value of x for which $\ln(2x - 1) = 3 - \frac{x}{2}$.

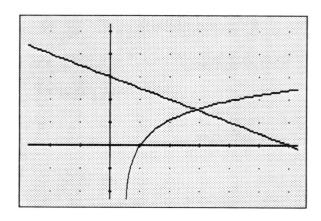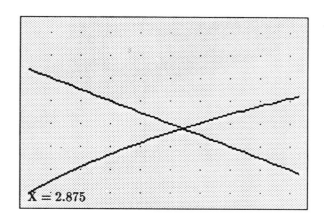

FIGURE 9.4. The graphs of $y = \ln(2x - 1)$ and $y = 3 - \frac{x}{2}$. Zoomed three times on the right.

How good an approximation is 2.875? It is easy enough to check how close $\ln(2x - 1)$ and $3 - \frac{x}{2}$ are when $x = 2.875$:

2.875 $\boxed{\rightarrow}$ $\boxed{\text{ALPHA}}$ $\boxed{\text{X}}$

$\boxed{\ln}$ $\boxed{(}$ 2 $\boxed{\text{ALPHA}}$ $\boxed{\text{X}}$ $\boxed{-}$ 1 $\boxed{)}$ $\boxed{\text{EXE}}$ displayed value: 1.55814

3 $\boxed{-}$ $\boxed{\text{ALPHA}}$ $\boxed{\text{X}}$ $\boxed{\div}$ 2 $\boxed{\text{EXE}}$ displayed value: 1.56250

DIRECTIONS: Show all work to each question right on these pages. For graphs, make a sketch based on the Casio's graphing screen display.

Use the zooming-in technique described in this experiment to approximate solution(s) to the given equations in Exercises 1 through 8. Express your answers to the nearest hundredth. Check with your instructor to see if he/she also wants you to turn in sketches of the graphs.

1. $4^{x+3} = 7^x$ Solution(s): _____

2. $e^x = 3 - 2e^{-x}$ Solution(s): _____

3. $1 + e^{2x} = \frac{4}{3}$ Solution(s): _____

4. $\ln(1 - x) = \ln 6 - \ln(x + 4)$ Solution(s): _____

5. $\log x = \ln x$ Solution(s): _____

6. $4 - x^2 = \ln x$ Solution(s): _____

7. $x^2 = 2^x$ Solution(s): _____

8. $\dfrac{e^x + e^{-x}}{e^x - e^{-x}} = 2$ Solution(s): _____

9. Graph the following four functions on the same set of coordinate axes:

$$y = \ln x \qquad y = \frac{1}{e}x \qquad y = \frac{1}{e}x + 1 \qquad y = \frac{1}{e}x - 1$$

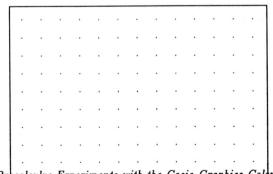

Precalculus Experiments with the Casio Graphics Calculator

Using the information in your graph, what can be said about the *number* of solutions to each of the following equations:

      a. $\ln x = \frac{1}{e}x$ How many solutions? _____

      b. $\ln x = \frac{1}{e}x + 1$ How many solutions? _____

      c. $\ln x = \frac{1}{e}x - 1$ How many solutions? _____

10. a. One way to find out how many years it will take for an investment of P dollars at 8.5% annual interest compounded quarterly to triple is to solve the following equation for t:

$$\left(1 + \tfrac{r}{n}\right)^{nt} = 3 \quad \text{where } r = 0.085 \text{ and } n = 4$$

$t =$ _____

b. Solve the problem in part a graphically by graphing $y = \left(1 + \dfrac{0.085}{4}\right)^{4x}$ and $y = 3$ and finding the abscissa of their intersection.

11.   When Michelle was presented with the task of approximating the solution to the equation, $e^{0.3 - x} = 1 - 2x^3$, she decided to graph the system:

$$y = e^{0.3 - x}$$
$$y = 1 - 2x^3$$

and search for the point of intersection. She got the graph below on her Casio fx-7500G calculator:

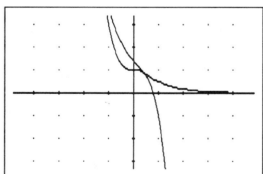

a. Why was Michelle's guess that the solution was between 0 and 1 a bad decision?

_____

_____

_____

b. Approximate the solution to the equation $e^{0.3 - x} = 1 - 2x^3$ using the graphical approach and a better range setting than Michelle did.

Introduction to Trigonometry

INTRODUCTION

1. Angles can be measured in a variety of ways. In most mathematics courses, we use three ways: decimal degrees, degrees-minutes-seconds, and radians. For example, all three of the following represent the same size angle: 25.5°, 25°30', 0.44506 radians. Conversion from one type of measurement to another can be done by textbook formula or directly on the Casio calculator.

2. Radian measure is especially useful in many formulas including arc length, area of a circular sector and angular velocity. The formula for arc length, s, in a circle of radius r and central angle θ, for example, is $s = r\theta$. The area of a circular sector is $A = \frac{1}{2}\theta r^2$ and the formula for angular speed is $\omega = \frac{\theta}{t}$. In each of these formulas, θ must be measured in radians.

PROCEDURES

Procedure 1. You will often want to be in radian mode. Recall that this is done by pressing MODE 5 EXE. The M Disp key will show you the present mode if you forgot to check it out when the calculator was turned on. The mode display should resemble:

```
* * * * MODE * * * *
sys mode :    RUN
cal mode :    COMP
    angle:    Rad
  display:    Fix 5

    Step     0
```

Now, to convert an angle measured in degrees to radian measure, we multiply by $\frac{\pi}{180}$. For example, 25.5° gets converted to radians as follows:

25.5 $\boxed{\times}$ $\boxed{\text{SHIFT}}$ $\boxed{\pi}$ $\boxed{\div}$ 180 $\boxed{\text{EXE}}$ Display: 0.44506

Another option is to use the $\boxed{\text{SHIFT}}$ $\boxed{\text{MODE}}$ combination. Remember, you must be in radian mode:

25.5 $\boxed{\text{SHIFT}}$ $\boxed{\text{MODE}}$ 4 $\boxed{\text{EXE}}$ Display: 0.44506

Think of " $\boxed{\text{SHIFT}}$ $\boxed{\text{MODE}}$ 4 $\boxed{\text{EXE}}$ " as a "temporary" degree mode.

To convert radians to degrees by multiplication, we multiply by $\frac{180}{\pi}$. For example, to convert 4.5 radians to degrees we could perform the following keystrokes:

4.5 $\boxed{\times}$ 180 $\boxed{\div}$ $\boxed{\text{SHIFT}}$ $\boxed{\pi}$ $\boxed{\text{EXE}}$ Display: 257.83101

Alternatively, we could enter degree mode ($\boxed{\text{MODE}}$ 4 $\boxed{\text{EXE}}$) and perform the following:

4.5 $\boxed{\text{SHIFT}}$ $\boxed{\text{MODE}}$ 5 $\boxed{\text{EXE}}$ Display: 257.83101

Now, we will use the $\boxed{\circ\;'\;''}$ key to convert an angle measure like 160° 30' 20" to decimal degrees as follows:

160 $\boxed{\circ\;'\;''}$ 30 $\boxed{\circ\;'\;''}$ 20 $\boxed{\circ\;'\;''}$ $\boxed{\text{EXE}}$ Display: 160.50556

To convert in the other direction, i.e. from decimal degrees to degrees-minutes-seconds, enter the number and then use the $\boxed{\text{SHIFT}}$ version of the $\boxed{\circ\;'\;''}$ key for the conversion. For example, to convert $-112.6°$ to degrees-minutes-seconds, press the following:

$\boxed{(-)}$ 112.6 $\boxed{\text{EXE}}$ $\boxed{\text{SHIFT}}$ $\boxed{\Longleftarrow}$ Display: $-112°36'00"$

Procedure 2.   If you had many problems where you had to find the arc length and the area of circular sectors, it would behoove you to use a program to minimize your keystrokes. Below, we have entered a program, P5, that will calculate the arc length and area for a sector.

Logic of what must be done:

1. The value of the circle's radius is stored in storage location R and the central angle should be entered and stored in storage location T.

2. Calculate and display the arc length.

3. Calculate and display the area.

Casio syntax

1. "RADIUS = " ? → R :
   "THETA = " ? → T :

2. "ARC LEN = " :
   T × R ◄

3. "AREA = " :
   .5 × T × R² ◄

Of course, the program only works as well as its input. That is, the values of θ input must be in radians, *not degrees*. A list of input and output for various radii and central angles appears below.

Radius	Angle	Input R	Input θ	Output Arc	Output Area
5"	0 .85	5	0.85	4.25 (in)	10.625 (in²)
2 cm	30°	2	$\frac{\pi}{6}$	1.04720 (cm)	1.04720 (cm²)
19 ft	225°	19	3.92699	74.61281 (ft)	708.82170 (ft²)

We conclude this experiment with an applied problem that involves the notion of angular speed.

A car is moving at the rate of 55 miles per hour and the diameter of each of its wheels is 2 feet. Find the number of revolutions per minute that the wheels are rotating and find the angular speed of the wheels.

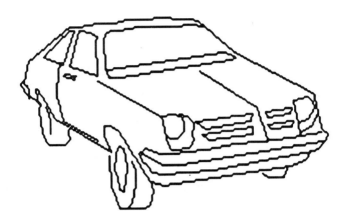

To find the number of revolutions per minute of the wheel, it needs to be noted that one revolution means a point on the wheel travels its circumference, $\pi \times 2$ ft. The rest of the solution below is a matter of handling units so that the final (uncanceled) units are rev/min:

$$\frac{1\ rev}{2\pi\ ft} \cdot \frac{5280\ ft}{1\ mile} \cdot \frac{55\ miles}{1\ hr} \cdot \frac{1\ hr}{60\ min} \approx 770.3\ \frac{rev}{min}$$

The angular speed of the wheels is expressed in radians per min:

$$770.3\ \frac{rev}{min} \cdot \frac{2\pi\ radians}{rev} \approx 4{,}840\ radians\ per\ minute.$$

DIRECTIONS: Show all work to each question right on these pages. For graphs, make a sketch based on the Casio's graphing screen display.

In questions 1 through 10, express each angle three ways: in radian measure, as decimal degrees, and in degrees-minutes-seconds format.

	Radians	Decimal Degrees	Degrees-Minutes-Seconds
1.	.57	_____	_____
2.	_____	211.8°	_____
3.	_____	_____	132° 35' 12"
4.	$3\pi/4$	_____	_____
5.	_____	_____	− 67° 42' 30"
6.	10.8	_____	_____
7.	_____	225.6°	_____
8.	− 2.9	_____	_____
9.	_____	_____	18° 18' 18"
10.	_____	123.45°	_____

In questions 11 through 13, use program P5 or a variation of it to find the requested information.

11. What is the length of the arc subtended by a central angle of 39° in a circle of radius 4 cm?

12. If a circular pizza is 18 inches in diameter and is cut into eight equal pieces:

      a. What is the area of each piece of pizza?

      b. What is the crust length of each piece of pizza?

13. What is the length of arc subtended by a central angle of 101° 56' 12" in a circle with a diameter of 19 feet?

14. A car is moving at a rate of 80 km/hr and the diameter of each wheel is 0.9 meters.

      a. Find the number of revolutions per minute that the wheels are rotating.

      b. Find the angular speed of the wheels (in radians per minute).

15. How long will it take a pulley rotating at 8 radians per second to make 50 complete revolutions?

Cumulative Review Exercises

1. a) The graph of $y = x^2 + 6$ does not appear on the graphics screen when the default range settings are in effect. Explain why not. _____

_____

b) Sketch the graph of $y = x^2 + 6$ and state the range setting used to get the sketch.

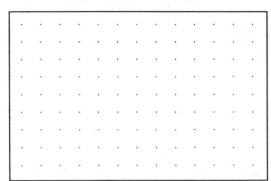

Range

Xmin:
  max:
  scl:
Ymin:
  max:
  scl:

2. Sketch a graph of $y = |4 - x^2|$. What is the function's domain and range?

Domain: _____

Range: _____

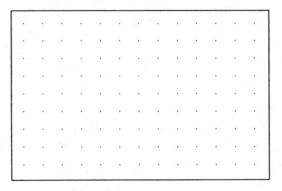

3. Consider the function $f(x) = 6x^4 + 7x^3 - 11x^2 - x + 2$.

a) *Describe* the process of how you would find the zeros of f using a graphical approach.

_____

_____

b) What are the real zeros of f approximated to the nearest hundredth?

c) What are the real zeros of f *exactly*?

4. Suppose $f(x) = \sqrt[3]{x-1} + 2$ and $g(x) = (x-2)^3 + 1$. Choosing suitable range settings, graph $y = x$, $y = g(x)$, and $y = f(x)$ on the same set of axes. Make a conclusion about f and g.

Range

Xmin:

  max:

  scl:

Ymin:

  max:

  scl:

Conclusion:

_____

_____

5. a) Make a sketch of the quadratic function $y = 5x^2 + 34x - 7$ and its axis of symmetry. State the range settings you had the Casio use. Also, state the vertex and the axis of symmetry.

Range

Xmin:

  max:

  scl:

Ymin:

  max:

  scl:

Axis of symmetry: _____

Coordinates of vertex: _____

b) Explain how this graph is related to the graph of $y = 5x^2 + 34x$.

c) Explain how the graph of $y = 5x^2 + 34x$ is related to the graph of $y = 5(x-1)^2 + 34(x-1)$.

6. Graph $y = \dfrac{5x}{x^2 - 4}$

          Domain: _____

          Range: _____

          Vertical Asymptotes: _____

          Horizontal Asymptotes: _____

          Intercepts: _____

          Symmetry: _____

Is this function even, odd, or neither? _____

7. Carefully examine the rational function $f(x) = \dfrac{2x^4 + 2x^3 - 7x^2 - 3x + 6}{3x^4 + 5x^3 + 3x^2 + 9x - 2}$. For parts a and b below, use the rational root theorem and other related algebraic tools to best answer the question.

a) For what real value(s) of x is the numerator zero?

b) For what real value(s) of x is the denominator zero? (HINT: There are two. Give one exactly and approximate the other.)

c) What is $f(-2)$ and what should the graph of $y = f(x)$ look like at $x = -2$?

d) What is the vertical asymptote (approximately)?

e) What is the horizontal asymptote?

f) Sketch the graph. (You may actually need to observe the graph under a variety of graph settings to get an accurate sketch.)

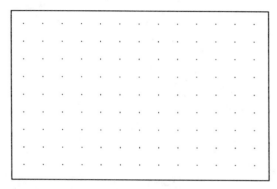

In questions 8 and 9, use the Casio to graphically approximate the solutions to each equation. Express your answer to the nearest hundredth.

8. $2 - x^2 = \ln x$ Solution(s): _____

9. $4x^3 - x^2 = e^x$ Solution(s): _____

In questions 10 through 18, perform each calculation on the Casio calculator. Set the calculation mode to be FIXed at 6 decimal places. Before you begin, assume the following values: $A = 2$, $B = -3$, $C = 5.6$, $D = \sqrt{3}$. If the calculator gives an error message, state what the problem is likely to be with given expression.

10. A + C/2B   10. _____

11. $D^2 - 4ABC$

11. _____

12. $\dfrac{A+6}{10C+19B+3}$

12. _____

13. $\sqrt{A} + \sqrt{B} + \sqrt{C} + \sqrt{D}$

13. _____

14. $\sqrt{A+B+C+D}$

14. _____

15. $\sqrt[3]{B}$

15. _____

16. $\log 5A$

16. _____

17. $\ln(A-3B)$

17. _____

18. e^D

18. _____

19. Convert each of the following radian measures to a) decimal degrees and b) degrees-minutes-seconds format.

Radian measure	Decimal Degrees	D-M-S
1.23	_____	_____
-2.19	_____	_____

20. Convert each of the following degree measures to radian measure.

81.9°   Radian equivalent: _____

$-210°50'30''$ Radian equivalent: _____

$\boxed{\text{Experiment \#11}}$

Evaluating Trigonometric Functions

INTRODUCTION

1. There are six trigonometric functions of an angle θ: sine, cosine, tangent, cotangent, secant, and cosecant. There definitions should be learned -- see your textbook.

   Notation is sloppy regarding the trigonometric functions. First, each of the six functions above has an abbreviation: sin, cos, tan, cot, sec, csc. Now, each of these functions has an argument and to be consistent we *should* place that argument in parentheses (supporting the notation of functions). However, instead of writing "sin(θ)" we often abbreviate it as sin θ.

   Various relationships exist among the trigonometric functions; these relationships are often called identities. For example, one reciprocal identity is $\sin \theta = \frac{1}{\csc \theta}$. Only the first three trigonometric functions are represented with keys on the Casio calculator; the other three must be obtained via their reciprocals.

2. The calculator is a blessing when it comes to evaluating the trigonometric functions of any angle. Prior to the electronic calculator, tables were used -- usually with four or five place accuracy. In that respect, they were not "the good ole days"!

   Trigonometric functions of some angles are undefined. For example, by its definition, the tangent of ninety degrees, tan 90°, involves division by zero.

3. There are six inverse trigonometric functions, $\sin^{-1}$, $\cos^{-1}$, $\tan^{-1}$, $\cot^{-1}$, $\sec^{-1}$, and $\csc^{-1}$ and again only the first three have key representations on the calculator.

PROCEDURES

Procedure 1.   The trigonometric functions of any angle can be evaluated, just make sure the calculator is in the correct mode to accept the angle. There is no mathematical limitation on the argument θ -- you can theoretically take the sine of any angle. However, there are practical limitations on the calculators and if θ is in degrees, then $-9 \times 10^9 \leq \theta \leq 9 \times 10^9$. If θ is expressed in radians, $-5\pi \times 10^7 \leq \theta \leq 5\pi \times 10^7$ is the restriction for using the $\boxed{\sin}$, $\boxed{\cos}$, and $\boxed{\tan}$ keys.

As long as the mode is respected, evaluating trigonometric functions is very straight forward as the next examples show.

To evaluate ...	Enter these keystrokes...	Display:
$\sin 204.8°$	$\boxed{\text{MODE}}$ 4 $\boxed{\text{EXE}}$ $\boxed{\sin}$ 204.8 $\boxed{\text{EXE}}$	-0.41945
$\cos 295° \, 43' \, 50"$	$\boxed{\text{MODE}}$ 4 $\boxed{\text{EXE}}$ $\boxed{\cos}$ 295 $\boxed{°\,'\,"}$ 43 $\boxed{°\,'\,"}$ 50 $\boxed{°\,'\,"}$ $\boxed{\text{EXE}}$	0.43414
$\tan\left(\frac{\pi}{3}\right)$	$\boxed{\text{MODE}}$ 5 $\boxed{\text{EXE}}$ $\boxed{\tan}$ $\boxed{(}$ $\boxed{\text{SHIFT}}$ $\boxed{\pi}$ $\boxed{\div}$ 3 $\boxed{)}$ $\boxed{\text{EXE}}$	1.73205
$\sec 110.9°$	$\boxed{\text{MODE}}$ 4 $\boxed{\text{EXE}}$ $\boxed{(}$ $\boxed{\cos}$ 110.9 $\boxed{)}$ $\boxed{x^{-1}}$ $\boxed{\text{EXE}}$	-2.80318
$2\sin 55° + \cos 63°$	$\boxed{\text{MODE}}$ 4 $\boxed{\text{EXE}}$ 2 $\boxed{\times}$ $\boxed{\sin}$ 55 $\boxed{+}$ $\boxed{\cos}$ 63 $\boxed{\text{EXE}}$	2.209229
$\sin^{-1} 0.6$	$\boxed{\text{MODE}}$ 4 $\boxed{\text{EXE}}$ $\boxed{\text{SHIFT}}$ $\boxed{\sin^{-1}}$ 0.6 $\boxed{\text{EXE}}$ $\boxed{\text{MODE}}$ 5 $\boxed{\text{EXE}}$ $\boxed{\text{SHIFT}}$ $\boxed{\sin^{-1}}$ 0.6 $\boxed{\text{EXE}}$	36.86990 (degrees) 0.64350 (radians)

Remember, regardless of the angle θ, the following range statements are always true:

$$-1 \leq \sin\theta \leq 1$$

$$-1 \leq \cos\theta \leq 1$$

<u>Procedure 2.</u> The tangent of each of 90°, 270°, 450°, − 90°, etc is undefined. The Casio returns the following error message when a trigonometric function evaluation is attempted that is undefined:

Ma ERROR

Step   3

Similar undefined "Ma ERROR" statements occur for sec 90°, $\sin^{-1}2.3$, and cot π.

<u>Procedure 3.</u> Trigonometric functions are periodic and not one-to-one. For example, there are infinitely many values of θ for which tan θ = 1. By restricting θ by 0° $\leq$ θ < 360°, there are *two* values of θ now for which tan θ = 1; one is in the first quadrant and the other is in the third quadrant. Here, you will need to follow your textbook's development closely. The notion of <u>reference angle</u> is crucial for finding that third quadrant angle because the calculator only returns the first quadrant angle (45°) when $\boxed{\text{SHIFT}}$ $\boxed{\tan^{-1}}$ 1 $\boxed{\text{EXE}}$ is entered. **Beware!** Sometimes the calculator returns an angle not in the desired interval.

Study carefully the values below and make sure *you* can come up with *all* the values even though your calculator cannot!

$\boxed{\text{Assume } 0° \leq \theta < 360° \text{ for the calculations below.}}$

Given ...	Then the angle θ must be in ...	The values of θ are ...
sin θ = 0.35	Quadrant 1 or	20.48732°
	Quadrant 2	159.51268°
cos θ = − 0.1234	Quadrant 2 or	97.08837°
	Quadrant 3	262.91163°
tan θ = − 2.8192	Quadrant 2 or	109.53013°
	Quadrant 4	289.53013°
sec θ = 12.6	Quadrant 1 or	85.44793°
	Quadrant 4	274.55207°

NOTES

Name _____   Experiment #11

Date _____   Exercise Sheet

DIRECTIONS: Show all work to each question right on these pages.

In questions 1 through 20, use the Casio to evaluate each of the trigonometric expressions, if possible. If the expression is undefined, write "UNDEFINED".

1. $\sin 79°$

_____

2. $\cos(-59° 32')$

_____

3. $\tan 12$

_____

4. $\sec 129°$

_____

5. $\cot(199° 62' 44'')$

_____

6. $\csc \frac{3\pi}{2}$

_____

7. $\sin 1$

_____

8. $\tan \frac{3\pi}{2}$

_____

9. $\cot \frac{3\pi}{2}$

_____

10. $\sin \frac{12\pi}{5}$

_____

11. $\sec(279° 10' 30'')$

_____

12. $\tan 12°$

_____

13. $\csc(-193.8°)$

_____

14. $2\sin 191.6° + 3\cos 81°$

_____

15. $\sin^2 \frac{7\pi}{6} + \cos^2 \frac{7\pi}{6}$

_____

16. $\frac{\sin 79°}{\cos 79°}$

_____

17. $\sqrt{\tan(2.6)}$

_____

18. $\tan \sqrt{2.6}$

_____

19. $\left(\sec(133°29' 42'')\right)\left(\cos(133°29' 42'')\right)$

_____

20. $\left[\sin(124°) - \cos(-35.4°)\right]^2$

_____

In questions 21 through 32, assume $0° \leq \theta < 360°$. Find the quadrants in which θ lies and find each value of θ expressing θ in degrees-minutes-seconds format. If there is no value of θ that satisfies the expression, write "UNDEFINED".

21. $\tan \theta = 1.5000$ Quadrants: _____ $\theta_1 =$ _____ $\theta_2 =$ _____

22. $\sin \theta = -0.3156$ Quadrants: _____ $\theta_1 =$ _____ $\theta_2 =$ _____

23. $\cos \theta = 0.8199$ Quadrants: _____ $\theta_1 =$ _____ $\theta_2 =$ _____

24. $\tan \theta = -1.5000$ Quadrants: _____ $\theta_1 =$ _____ $\theta_2 =$ _____

25. $\sin \theta = 0.7319$ Quadrants: _____ $\theta_1 =$ _____ $\theta_2 =$ _____

26. $\cot \theta = -2.3164$ Quadrants: _____ $\theta_1 =$ _____ $\theta_2 =$ _____

27. $\cos \theta = 1.5000$ Quadrants: _____ $\theta_1 =$ _____ $\theta_2 =$ _____

28. $\cos \theta = 0.3196$ Quadrants: _____ $\theta_1 =$ _____ $\theta_2 =$ _____

29. $\cos \theta = -0.2938$ Quadrants: _____ $\theta_1 =$ _____ $\theta_2 =$ _____

30. $\sec \theta = 3.124$ Quadrants: _____ $\theta_1 =$ _____ $\theta_2 =$ _____

31. $\csc \theta = -2.0199$ Quadrants: _____ $\theta_1 =$ _____ $\theta_2 =$ _____

32. $\cos \theta = -1.0000$ Quadrants: _____ $\theta_1 =$ _____ $\theta_2 =$ _____

33. If θ is an acute angle in a right triangle and it is known that $\sin \theta = 0.2911$, find θ and all six trigonometric functions of θ.

$\theta =$ $\sin \theta =$ $\cos \theta =$ $\tan \theta =$ $\cot \theta =$

$\sec \theta =$ $\csc \theta =$

2. Have user input the values of A, B, and C and store them in those memory locations.

2. ”A = ” ? → A :
   ”B = ” ? → B :[14]
   ”C = ” ? → C :

3. Input the range settings for the Casio. Since the period is $2\pi/B$, one-fourth of that, $\pi/2B$, will make a suitable x-axis scale. C/B and $(2\pi + C)/B$ are endpoints of one cycle of the shifted curve.

3. Range C ÷ B, $(2\pi+C)$ ÷ B, π ÷ 2B,
   − Abs(A), Abs(A), 1 :

4. Use labels for the two basic parts. Label 1 is for graphing the sine curve and label 2 is for graphing the cosine curve.

4. D ≠ 1 ⇒ Goto 2 :
   Lbl 1 : Graph Y = A × sin (BX + C) ◀
   Lbl 2 : Graph Y = A × cos (BX + C) ◀

The remainder of this experiment provides you with sample output from running this program (⌈Prog⌉ 6 ⌈EXE⌉) six times. When you run it, you are urged to keep checking the ⌈Range⌉ screen so that the graph is viewed from the correct perspective.

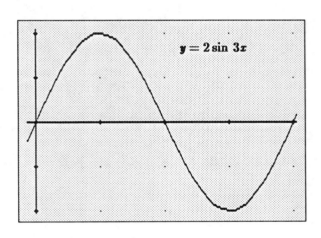

FIGURE 12.7. A = 2, B = 3, C = 0, D = 1

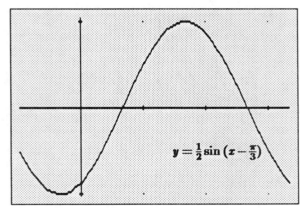

FIGURE 12.8. A = $\frac{1}{2}$, B = 1, C = $-\frac{\pi}{3}$, D = 1

---

[14]For the sake of simplicity, we assume that B is always positive. The program would have to be adapted to allow for negative B values.

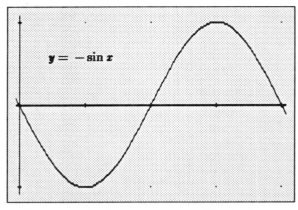

FIGURE 12.9. A = −1, B = 1, C = 0, D = 1

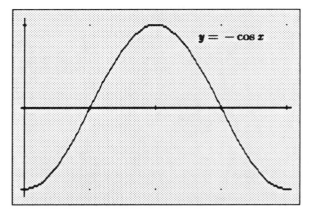

FIGURE 12.10. A = −1, B = 1, C = 0, D = 2

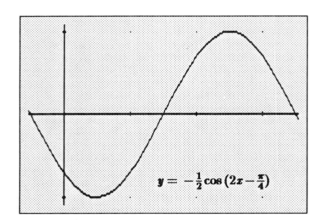

FIGURE 12.11. A = −$\frac{1}{2}$, B = 2, C = −$\frac{\pi}{4}$, D = 2

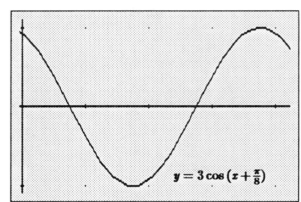

FIGURE 12.12. A = 3, B = 1, C = $\frac{\pi}{8}$, D = 2

Name _____   Experiment #12

Date _____   Exercise Sheet

DIRECTIONS: Show all work to each question right on these pages. For graphs, make a sketch based on the Casio's graphing screen display.

In questions 1 through 6, make a sketch of the graph of each function. Also, state its amplitude, period, and phase shift.

1. $y = 2\sin(4x)$ Amp: _____

Period: _____ Phase Shift: _____

2. $y = 4\sin\left(3x - \frac{\pi}{4}\right)$ Amp: _____

Period: _____ Phase Shift: _____

3. $y = 2\cos\left(\frac{1}{2}x\right)$ Amp: _____

Period: _____ Phase Shift: _____

4. $y = 4\cos\left(3x - \frac{\pi}{4}\right)$ Amp: _____

Period: _____ Phase Shift: _____

5. $y = -2\sin\left(3x - \frac{\pi}{4}\right)$ Amp: _____

Period: _____ Phase Shift: _____

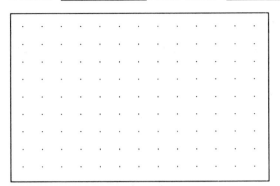

6. $y = -0.5\cos\left(3x + \frac{\pi}{2}\right)$ Amp: _____

Period: _____ Phase Shift: _____

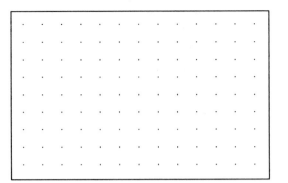

7. Graph both of the following functions on the same set of coordinate axes: $y = 2\sin 3x$ and $y = 4 + 2\sin 3x$. What can be said about $y = A\sin(Bx + C)$ and $y = G + A\sin(Bx + C)$?

8. a. Graph $y = \sin x$ and $y = \cos\left(x - \frac{\pi}{2}\right)$ on the same set of coordinate axes.

   b. Graph $y = \cos x$ and $y = -\sin\left(x - \frac{\pi}{2}\right)$ on the same set of coordinate axes.

   c. Can you see that a sine curve is a displaced cosine curve and that a cosine curve is a negated displaced sine curve?

In questions 9 through 11, match the algebraic representation of the function with its graphical representation by placing the label of the function below the figure. The three functions are $y = 2\sin x$, $y = 2\sin\left(x + \frac{\pi}{8}\right)$, and $y = 2\sin\left(x - \frac{\pi}{8}\right)$. In each case the following range settings are in place: Xmin $= -0.1$, Xmax $= 6.4$, Xscl $= 1$, Ymin $= -3$, Ymax $= 3$, Yscl $= 1$.

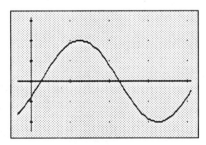  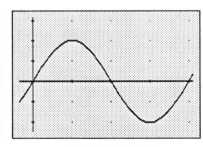

    FIGURE 12.13.   **FIGURE 12.14.**   **FIGURE 12.15.**

9. $y =$ _____   10. $y =$ _____   11. $y =$ _____

Graphs of Other Trigonometric Functions

INTRODUCTION

1. The graph of the tangent function differs considerably from the graphs of sine or cosine because there are values for which the tangent is undefined. The graphical interpretation of this is the appearance of vertical asymptotes at $x = \frac{n\pi}{2}$ (where n is any odd integer).

2. The cotangent, secant, and cosecant functions also have vertical asymptotes. After all, they are all defined as the reciprocal of functions which have zeros. So, where sine is zero, cosecant is undefined; where cosine is zero, secant is undefined.

3. The range of the tangent function is **R** and it makes no sense to discuss amplitude. (We can say the amplitude is undefined.) Similar statements are true for cotangent. The ranges for the secant and cosecant functions are $y \geq 1$ or $y \leq -1$. The tangent and cotangent functions have periods of π; the secant and cosecant functions have periods of 2π.

PROCEDURES

Procedure 1. To graph the trigonometric functions in this experiment, we will proceed as before with graphs: enter the range settings on the Casio and then use the $\boxed{\text{Graph}}$ key to plot it. Use the ranges and periods as a guide in entering the range settings. In Figure 13.1 below we graph the tangent function $y = \tan x$ by entering $\boxed{\text{Graph}}$ $\boxed{\text{tan}}$ $\boxed{\text{ALPHA}}$ $\boxed{\text{X}}$.

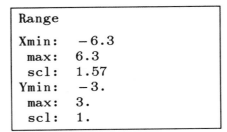

```
Range
Xmin:  -6.3
 max:   6.3
 scl:   1.57
Ymin:  -3.
 max:   3.
 scl:   1.
```

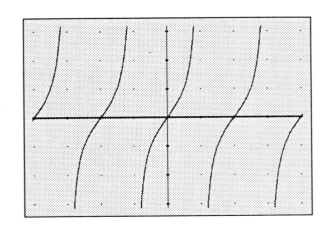

Period: π

Domain: $x \neq \frac{n\pi}{2}$ where is n is an odd

integer.

FIGURE 13.1. $y = \tan x$

The graphs of the remaining three trigonometric functions appear below. To create them, we entered the following sequence of keystrokes. [In each case, the range settings are identical to those of Figure 13.1.

For $y = \cot x$ [Graph] [( ] [tan] [ALPHA] [X] [ )] [x^{-1}] [EXE]

For $y = \sec x$ [Graph] [( ] [cos] [ALPHA] [X] [ )] [x^{-1}] [EXE]

For $y = \csc x$ [Graph] [( ] [sin] [ALPHA] [X] [ )] [x^{-1}] [EXE]

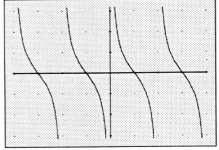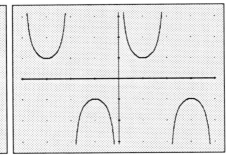

FIGURE 13.2. $y = \cot x$

Domain: $x \neq n\pi$ ($n =$ even integer)

Range: $\mathbb{R}$

Period: π

FIGURE 13.3. $y = \sec x$

Domain: $x \neq \frac{n\pi}{2}$ ($n =$ odd integer)

Range: $y \geq 1$ or $y \leq -1$

Period: 2π

FIGURE 13.4. $y = \csc x$

Domain: $x \neq n\pi$ ($n =$ even integer)

Range: $y \geq 1$ or $y \leq -1$

Period: 2π

In the exercises, you will have practice graphing variations of these functions.

Name _____   **Experiment #13**

Date _____   **Exercise Sheet**

DIRECTIONS: Show all work to each question right on these pages. For graphs, make a sketch based on the Casio's graphing screen display.

In questions 1 through 8, graph each of the given functions.

1. $y = \tan\left(x - \frac{\pi}{4}\right)$

2. $y = \tan 3x$

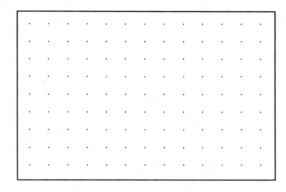

3. $y = 3\cot x$

4. $y = \sec\frac{1}{2}x$

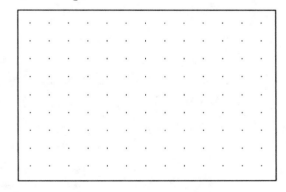

5. $y = \csc\left(2x + \frac{\pi}{4}\right)$

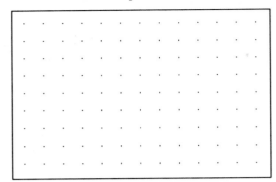

6. $y = \frac{1}{2}\sec\left(3x - \frac{\pi}{8}\right)$

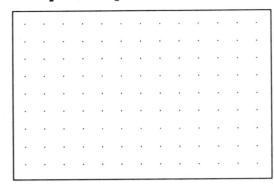

7. $y = \cot\left(x - \frac{\pi}{4}\right)$

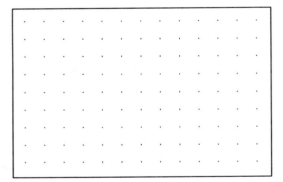

8. $y = \csc \pi x$

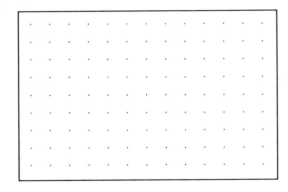

9. A technique called "addition of ordinates" can be used to manually graph a function such as $y = \sin x + 2\cos x$ but of course the Casio does a very nice job of graphing it, too! Using a suitable range setting, graph this function in the space below.

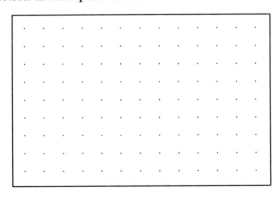

Inverse Trigonometric Functions

INTRODUCTION

1. The inverse sine function, denoted $y = \sin^{-1}x$ or $y = \arcsin x$, means that y is the angle (or number) whose sine is x. It has domain $-1 \leq x \leq 1$ and range $-\frac{\pi}{2} \leq y \leq \frac{\pi}{2}$. The equations $y = \sin^{-1}x$ and $x = \sin y$ are identical provided the domains are restricted to provide a one-to-one relationship.

2. Similarly, $y = \cos^{-1}x$ is the inverse of $y = \cos x$. Its domain is $-1 \leq x \leq 1$ and its range is $0 \leq y \leq \pi$. For $y = \tan x$, the domain is $\mathbb{R}$ and its range is $-\frac{\pi}{2} < y < \frac{\pi}{2}$.

3. The graph of $y = \sin^{-1}x$ is the mirror image of the graph of $y = \sin x$ about the line $y = x$ since they are inverse functions of one another.

PROCEDURES

Procedure 1. We have already accessed the inverse trigonometric function keys on the Casio so evaluating an inverse trigonometric function is not new. To find the angle whose sine is -0.5, for example, key in the following:

For the answer in radians:

         MODE 5 EXE

         SHIFT sin⁻¹ (−) 0.5 EXE   Displayed value: -0.52360

For the answer in degrees:

         MODE 4 EXE

         SHIFT sin⁻¹ (−) 0.5 EXE   Displayed value: -30.00000

The functions $\cot^{-1}$, $\sec^{-1}$, and $\csc^{-1}$ are used much less than the other three. Since there are no keys for these three functions on the calculator, to evaluate one, such as $\cot^{-1}(2)$, for example, you will have to realize that this is identical to the evaluation $\tan^{-1}(\frac{1}{2})$.

For the answer to $cot^{-1}(2)$ in radians:

      $\boxed{\text{MODE}}$ 5 $\boxed{\text{EXE}}$

      $\boxed{\text{SHIFT}}$ $\boxed{\tan^{-1}}$ $\boxed{(}$ 2 $\boxed{)}$ $\boxed{x^{-1}}$ $\boxed{\text{EXE}}$ Displayed value: 0.46365

For the answer to $sec^{-1}(4)$ in degrees:

      $\boxed{\text{MODE}}$ 4 $\boxed{\text{EXE}}$

      $\boxed{\text{SHIFT}}$ $\boxed{\cos^{-1}}$ $\boxed{(}$ 4 $\boxed{)}$ $\boxed{x^{-1}}$ $\boxed{\text{EXE}}$ Displayed value: 75.52249

If you attempt to ask the calculator for an impossibility, such as $\sin^{-1} 2$, a "Ma ERROR" (mathematical error) message is returned.

Procedure 2.   The graphs of three inverse trigonometric functions are presented in the following three figures. In each case note the accompanying range settings were entered to accommodate for the function's domain and range as stated above.

Range	
Xmin:	-1.5
max:	1.5
scl:	0.5
Ymin:	$-2.$
max:	2.
scl:	0.5

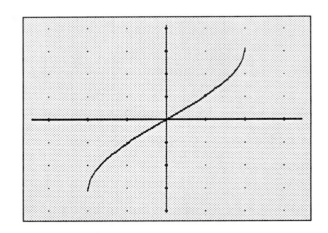

FIGURE 14.1. $y = \sin^{-1}x$

```
Range
Xmin:   -1.5
 max:    1.5
 scl:    0.5
Ymin:   -0.2
 max:    3.2
 scl:    0.5
```

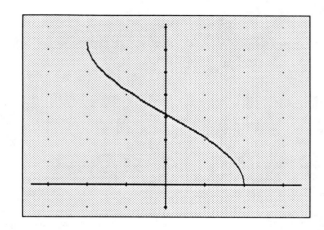

FIGURE 14.2. $y = \cos^{-1}x$.

```
Range
Xmin:   -4.7
 max:    4.7
 scl:    1.
Ymin:   -3.1
 max:    3.1
 scl:    1.
```

Note the horizontal asymptotes at

$x = -\frac{\pi}{2}$ and $x = \frac{\pi}{2}$.

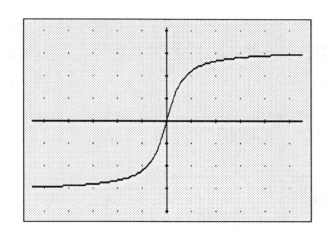

FIGURE 14.3. $y = \tan^{-1}x$

We conclude this experiment with some examples of using parentheses to enter complicated combinations of trigonometric and inverse trigonometric functions. For all of these calculations, be sure the calculator is in radian mode ([MODE] 5 [EXE]).

Mathematical expression	Keystrokes	Displayed value
$\tan\left(\arccos \frac{2}{3}\right)$	[tan] [(] [SHIFT] [cos⁻¹] [(] 2 [÷] 3 [)] [)] [EXE]	1.11803
$\cos\left[\arcsin\left(-\frac{3}{5}\right)\right]$	[cos] [(] [SHIFT] [sin⁻¹] [(] [(-)] 3 [÷] 5 [)] [)] [EXE]	0.80000
$\tan\left[\tan^{-1}(-3.5)\right]$	[tan] [(] [SHIFT] [tan⁻¹] [(] [(-)] 3.5 [)] [)] [EXE]	-3.50000
$\sin^{-1}\left(\sin \frac{5\pi}{3}\right)$	[SHIFT] [(] [sin] [(] 5 [SHIFT] [π] [÷] 3 [)] [)] [EXE]	-1.04720

Name _____   Experiment #14

Date _____   Exercise Sheet

DIRECTIONS: Show all work to each question right on these pages. For graphs, make a sketch based on the Casio's graphing screen display.

In questions 1 through 4, evaluate each expression. Express your answer in radians. If the expression is not defined, write "undefined" and state why.

1. $\sin^{-1}\left(-\frac{\sqrt{3}}{2}\right)$

   _____

2. $\cos^{-1}\left(\frac{3}{5}\right)$

   _____

3. $\tan^{-1}(-2.9012)$

   _____

4. $\cos^{-1}\left(\frac{\pi}{2}\right)$

   _____

In questions 5 through 8, evaluate each expression. Express your answer in decimal degrees. If the expression is not defined, write "undefined" and state why.

1. $\cos^{-1}\left(-\frac{\sqrt{3}}{2}\right)$

   _____

2. $\sin^{-1}\left(-\frac{4}{5}\right)$

   _____

3. $\sec^{-1}(-1.987)$

   _____

4. $\cot^{-1}\left(\frac{\pi}{2}\right)$

   _____

In questions 9 and 10, use the Casio to make a sketch of the graph of the given function.

5. $y = \sin^{-1}(2x)$

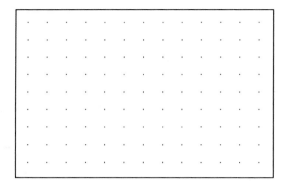

6. $y = \cot^{-1}x$

In questions 11 through 15, evaluate each expression.

11. $\tan\left[\tan^{-1}\left(-\frac{4}{3}\right)\right]$

_____

12. $\sin\left[\sin^{-1}\left(-\frac{\sqrt{3}}{2}\right)\right]$

_____

13. $\sec\left[\cos^{-1}\left(-\frac{1}{3}\right)\right]$

_____

14. $\cos\left[\sin^{-1}\left(-\frac{\sqrt{3}}{2}\right)\right]$

_____

15. $\cot\left[\cot^{-1}(-3.0129)\right]$

_____

Approximating Solutions to Trigonometric Equations

INTRODUCTION

1. Although a variety of important algebraic techniques should be studied for solving trigonometric equations, we can also rely on a graphical approach to approximate solutions. In fact, the approximating technique we will study here, is an excellent way to check your algebraically found solutions. Recall, that finding solutions to the equation $3\tan^2 x - 1 = 0$ is equivalent to finding the x-intercepts on the graph of $y = 3\tan^2 x - 1$.

2. We will be doing most of the work in this experiment on the Casio graphing screen and the calculator should be in radian mode before you begin. $\left[\ \boxed{\text{MODE}}\ 5\ \boxed{\text{EXE}}\ \right]$

PROCEDURES

Procedure 1. The technique is virtually identical to the one introduced in Experiment #9. Consider the equation $3\tan^2 x - 1 = 0$. For our first example, we will want to solve that equation for x provided x is in the interval $0 \leq x < 2\pi$. We will graph the function $y = 3\tan^2 x - 1$ and approximate its x-intercept. Then we will zoom-in on that point using the $\boxed{\text{Trace}}$ and $\boxed{\text{SHIFT}}$ $\boxed{\text{X}}$ keys. That process gets continued until we are confident we are "close" (within a predetermined tolerance) of the actual value.[15]

In Figure 15.1, we graph the function and, as usual, display the Casio range settings along with the graph.

---

[15]Alternatively, we could graph the two equations $y = 3\tan^2 x$ and $y = 1$ and zoom in on their point of intersection. We will choose the intercept approach in this experiment.

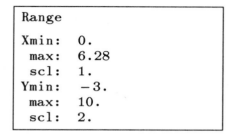

Range
Xmin: 0.
 max: 6.28
 scl: 1.
Ymin: -3.
 max: 10.
 scl: 2.

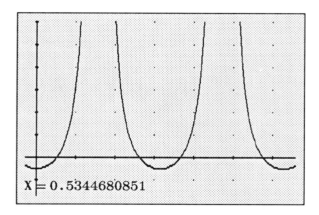

X = 0.5344680851

In $[0, 2\pi)$ there are four solutions to $\tan^2 x - 1 = 0$ because there are four x-intercepts in the graph of

$y = 3\tan^2 x - 1$ in that interval.

FIGURE 15.1. $y = 3\tan^2 x - 1$.

The [Trace] key is used to display the approximate x-intercept.

Now if we repeat the three-step process of graphing, tracing, and zooming five times, we will find that, to the nearest hundredth, a solution to the equation $3\tan^2 x - 1 = 0$ is 0.52. Figure 15.2 shows the result of the fifth such graphing.

Range
Xmin: 0.329867026
 max: 0.722367026
 scl: 1.
Ymin: -0.39217098
 max: 0.420362902
 scl: 2.

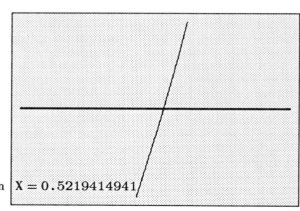

X = 0.5219414941

The values above were automatically calculated by the Casio when we used the [Trace] and [SHIFT] [x] combination five times.

FIGURE 15.2. $y = 3\tan^2 x - 1$ [Trace]d and zoom-in five times.

It can be shown that an *exact* solution to this equation is $\frac{\pi}{6}$. In exercise 1, the reader is asked to approximate the other three solutions of $3\tan^2 x - 1 = 0$ in $[0, 2\pi)$.

We reemphasize here some very important statements concerning the process we are using. It should be noted that all five of the following statements are equivalent and the reader should be familiar enough with the vocabulary of mathematics to understand these equivalences:

"The x-intercept of the graph of $y = 3\tan^2 x - 1$ is approximately (0.52, 0)."

"0.52 is an approximate solution to the equation $3\tan^2 x - 1 = 0$"

" 0.52 is an approximate zero of $3\tan^2 x - 1$ "

" If $f(x) = 3\tan^2 x - 1$, then $f(0.52) \approx 0$"

"The graphs of $y = 3\tan^2 x$ and $y = 1$, intersect at a point whose abscissa is about 0.52."

Procedure 2. Next, we want to solve (approximately) the equation $\cot x\ \cos^2 x = 2\cot x$. We begin by graphing the function $y = \cot x \cos^2 x - 2\cot x$ in Figure 15.3 below.

Range

Xmin: 0.
 max: 6.28
 scl: 1.
Ymin: −6.
 max: 6.
 scl: 2.

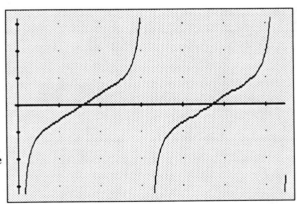

There appears to be two solutions to $\cot x\ \cos^2 x = 2\cot x$ in $[0, 2\pi)$ because are two x-intercepts to the graph of $y = \cot x\ \cos^2 x - 2\cot x$ in $[0, 2\pi)$.

FIGURE 15.3. $y = \cot x\ \cos^2 x - 2\cot x$.

After two trace/zoom combinations, we obtain the approximation 1.57 as is shown by Figure 15.4.

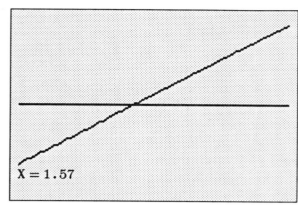

FIGURE 15.4. $y = \cot x \, \cos^2 x - 2 \cot x$ zoomed-in to show

an approximation for the x-intercept of 1.57.

In exercise 2, the reader is asked to approximate the other solution in the interval $[0, 2\pi)$.

<u>Procedure 3.</u> We conclude this experiment with one last example: find an approximate solution to the equation $\tan^2 x - 2 \tan x - 3 = 0$ on $[0, 2\pi)$.

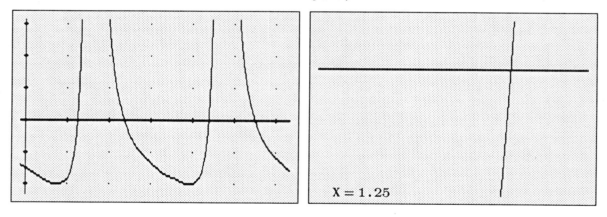

FIGURE 15.5. $y = \tan^2 x - 2 \tan x - 3$. The graph on the right is the result of zooming in twice.

It is interesting to point out that, had you used the traditional methods of finding a solution to this equation, you would have found that the *exact* solution occurs at $\tan^{-1} 3 \approx 1.2409$. The other three solutions to $y = \tan^2 x - 2 \tan x - 3$ in $[0, 2\pi)$ will be found in the exercises.

DIRECTIONS: Show all work to each question right on these pages.

1.   a. Use the techniques of your textbook to solve $3\tan^2 x - 1 = 0$.

   $x =$ _____

   b. We found one of those solutions to be approximately 0.52. Use the graphics calculator to approximate the other three solutions in the interval $0 \le x < 2\pi$ to the nearest hundredth. Compare your results with your answers in part a.

   _____   _____   _____

2.   a. Use the techniques of your textbook to solve $\cot x \cos^2 x - 2\cot x = 0$.

   $x =$ _____

   b. We found one of those solutions to be approximately 1.57. Use the graphics calculator to find the other solution in the interval $0 \le x < 2\pi$ to the nearest hundredth.

   _____

3.   a. Use the techniques of your textbook to solve $\tan^2 x - 2\tan x - 3 = 0$.

   $x =$ _____

   b. We found one solution to be approximately 1.25. In Figure 15.4 we can see that there are three other solutions in $[0, 2\pi)$. Use the graphics calculator to approximate them to two decimal places.

   _____   _____   _____

In questions 4 through 7, approximate all the solutions to the given equation in the interval $0 \leq x < 2\pi$. Round answers to the nearest hundredth.

4. $12 \sin^2 x - 13 \sin x + 3 = 0.$

$x =$ _____

5. $\csc^2 x - 2 = 0$

$x =$ _____

6. $\sin 2x = -\dfrac{\sqrt{3}}{2}$

$x =$ _____

7. $\dfrac{1 - \cos x}{1 + \cos x} = 0$

$x =$ _____

Solving Systems of Equations Graphically

INTRODUCTION

1. An equation in two variables, such as $x + y = 5$, has infinitely many solutions -- each solution being a pair of real numbers (x, y). Similarly the equation $x - y = -1$ has infinitely many solutions. Together, the two equations make up a <u>system of equations</u>. That system of equations may have a unique solution and one way to determine the solution is by graphing each equation.

2. If the graphs of both equations are straight lines, then one of three possibilities can occur: the lines could be parallel, in which case the system of equations has no solution; they may be coincidental (the same line), in which case there are infinitely many solutions to the system of equations; or the straight lines may meet in exactly one point, in which case the system of equations has a unique solution.

3. <u>Non-linear systems</u>, such as

$$x + y = 4$$
$$x^2 + y^2 = 4x$$

may have more than one unique solution. To graph an equation such as $x^2 + y^2 = 4x$ on the Casio graphics calculator, we must solve for y and graph the two functions, $y = \sqrt{4x - x^2}$ and $y = -\sqrt{4x - x^2}$.

PROCEDURES

<u>Procedure 1.</u> First, we will graph the system of linear equations in two variables:

$$x + y = 5$$
$$x - y = -1$$

We could do this by separately graphing each equation but we prefer to use the approach of program #P4 first seen in Experiment #9. Program P4 is just a convenient way of graphing two functions and it is composed of just two lines:[16]

$$\text{Graph Y} = 5 - X :$$
$$\text{Graph Y} = X + 1$$

Using the default range settings, we can just barely see that there is a point of intersection; to clarify the situation, we zoom-out after tracing as close to the point of intersection as we can, by pressing $\boxed{\text{SHIFT}} \boxed{\div}$. See below.

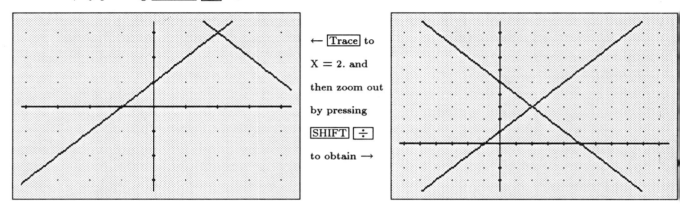

← $\boxed{\text{Trace}}$ to

X = 2. and

then zoom out

by pressing

$\boxed{\text{SHIFT}} \boxed{\div}$

to obtain →

FIGURE 16.1. $y = 5 - x$ and $y = x + 1$. The unique solution is (2, 3).

Of course, if the lines are parallel, there will be no point of intersection. In exercise 5, the reader will apply the above technique to observe parallel lines. Such a system of linear equations is called *inconsistent*.

Procedure 2. In non-linear systems, there may be more than one solution. For a first non-linear
    system consider:

$$y = \ln x$$
$$x + y = 1$$

The two equations are graphed in Figure 16.2.

---

[16]The reason for the program approach, rather than to separately enter and graph each function, has to do with the Casio's "replay" capability. We want to trace and zoom. When we zoom, we want both functions' graphs to appear and the program approach does this for us.

The two curves appear to intersect in exactly one point and in this case it is unnecessary to zoom in or out to see that the point of intersection is $(1, 0)$. Thus, $x = 1$ and $y = 0$ is the only solution to the system of two equations.

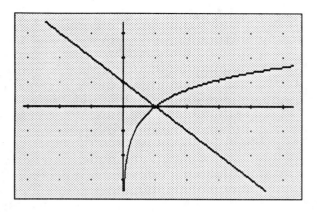

FIGURE 16.2. $y = \ln x$ and $y = 1 - x$ has one solution: $x = 1$ and $y = 0$.

Procedure 3. Finally, we graph the non-linear system of equations

$$x^2 + y^2 = 4x$$
$$x + y = 4$$

First, solve $x^2 + y^2 = 4x$ for y because the Casio will only graph *functions*. So, we graph **three** equations:

$$y = \sqrt{4x - x^2}$$

$$y = -\sqrt{4x - x^2}$$

$$y = 4 - x$$

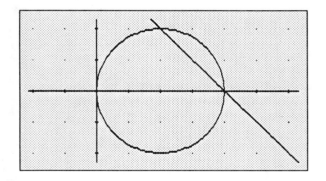

FIGURE 16.3. $x^2 + y^2 = 4x$ and $x + y = 4$. Two solutions: at $(4, 0)$ and at $(2, 2)$.

We conclude this experiment with two final examples: the first is a system of equations whose graphs are a parabola and a circle; the second is a non-linear system whose solutions include irrational numbers and we must approximate them using the trace and zoom features of the calculator.

Consider the system:

$$y = x^2$$
$$x^2 + (y - 1)^2 = 1$$

To graph it we need to adapt program P4 by entering the following keystrokes:

$\boxed{\text{Graph}}\ \boxed{\sqrt{\ }}\ \boxed{(\ }\ 1\ \boxed{-}\ \boxed{\text{ALPHA}}\ \boxed{\text{X}}\ \boxed{x^2}\ \boxed{)}\ \boxed{+}\ 1\ \boxed{\text{EXE}}$

$\boxed{\text{Graph}}\ \boxed{(-)}\ \boxed{\sqrt{\ }}\ \boxed{(\ }\ 1\ \boxed{-}\ \boxed{\text{ALPHA}}\ \boxed{\text{X}}\ \boxed{x^2}\ \boxed{)}\ \boxed{+}\ 1\ \boxed{\text{EXE}}$

$\boxed{\text{Graph}}\ \boxed{\text{ALPHA}}\ \boxed{\text{X}}\ \boxed{x^2}\ \boxed{\text{EXE}}$

Range	
Xmin:	$-2.$
max:	$2.$
scl:	$1.$
Ymin:	-0.25
max:	2.5
scl:	$1.$

The three solutions occur at

$(-1, 1)$, $(0, 0)$ and $(1, 1)$.

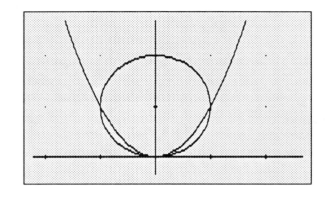

FIGURE 16.4. $y = x^2$ and $x^2 + (y-1)^2 = 1$

Finally we consider a system where we approximate the solutions:

$$y = 4 - x^2$$

$$y = \frac{4}{x^2}$$

Range	
Xmin:	-4.7
max:	4.7
scl:	$1.$
Ymin:	$-2.$
max:	5.5
scl:	$1.$

The $\boxed{\text{Trace}}$ key helps us approximate the x-values at ± 1.4 and both of the y-values at 2.04. You may want to zoom in once or twice by using $\boxed{\text{SHIFT}}\ \boxed{\text{X}}$.

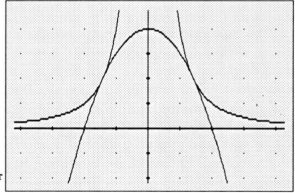

FIGURE 16.5. $y = 4 - x^2$ and $y = \frac{4}{x^2}$

In exercise 8, you are asked to compare the approximate with the exact results.

Name _____   **Experiment #16**

Date _____   **Exercise Sheet**

DIRECTIONS: Show all work to each question right on these pages. For graphs, make a sketch based on the Casio's graphing screen display.

In questions 1 through 4, use the traditional techniques of your textbook to solve the given system of equations. Then, check your results by using the Casio graphics calculator to verify your solutions. In each case, provide a sketch.

1. $y = 2x - 4$

   $x + y = 5$

   Solution: _____

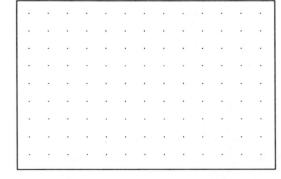

2. $y = 2x - 1$

   $2x + y = 9$

   Solution: _____

3. $x + y = -8$

   $2x - y = 2$

   Solution: _____

4. $x^2 + y^2 = 25$

   $y = x^2 - 5$

   Solutions: _____

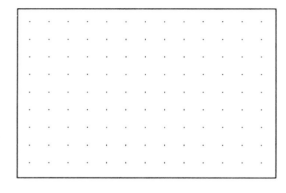

5. Solve the system $x - y = 3$ and $y = x$ and verify your answer on the Casio. In your own words, state the characteristics of this system .

   _____

   _____

6. Solve the system $x + y = 3$ and $y = \dfrac{6 - 2x}{2}$ and verify your answer on the Casio. In your own words, state the characteristics of this system .

   _____

   _____

7. Solve the system $y = x^2 - 2x + 1$ and $x + y = 3$ using the traditional, algebraic techniques learned in class and check the result with the Casio's graphs. Supply a sketch of the graphs below.

   Solution(s): _____

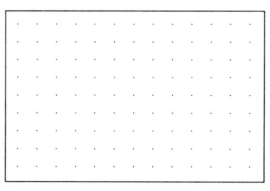

8. Find the *exact* solution(s) to the system $y = 4 - x^2$ and $y = \dfrac{4}{x^2}$. Compare those results with the results of Figure 16.5.

   Exact solution(s): _____

Determinants and Cramer's Rule

INTRODUCTION

1. Given a 2-by-2 matrix $\mathbf{A} = \begin{bmatrix} a_{11} & a_{12} \\ a_{21} & a_{22} \end{bmatrix}$, the <u>determinant of A</u>, denoted by $|\mathbf{A}|$ or by $\det(\mathbf{A})$, is

given by $\begin{vmatrix} a_{11} & a_{12} \\ a_{21} & a_{22} \end{vmatrix} = a_{11}a_{22} - a_{21}a_{12}.$

2. Given a 3-by-3 matrix $\mathbf{B} = \begin{bmatrix} b_{11} & b_{12} & b_{13} \\ b_{21} & b_{22} & b_{23} \\ b_{31} & b_{32} & b_{33} \end{bmatrix}$ the <u>determinant of B</u>, is given by:

$$\begin{vmatrix} b_{11} & b_{12} & b_{13} \\ b_{21} & b_{22} & b_{23} \\ b_{31} & b_{32} & b_{33} \end{vmatrix} = b_{11}b_{22}b_{33} + b_{12}b_{23}b_{31} + b_{13}b_{21}b_{32} - b_{31}b_{22}b_{13} - b_{32}b_{23}b_{11} - b_{33}b_{21}b_{12}$$

In this experiment we will confine our determinants to be either 2-by-2 or 3-by-3 and we will write a program to evaluate these determinants.

3. <u>Cramer's Rule</u> is an application of determinants to solving square (2 equations in 2 unknowns, 3 equations in 3 unknowns, etc) systems. Basically, the solutions to the general 2-by-2 system

$$a_{11}x + a_{12}y = c_1$$
$$a_{21}x + a_{22}y = c_2$$

are given by:

$$x = \frac{\begin{vmatrix} c_1 & a_{12} \\ c_2 & a_{22} \end{vmatrix}}{\begin{vmatrix} a_{11} & a_{12} \\ a_{21} & a_{22} \end{vmatrix}} \quad \text{and} \quad y = \frac{\begin{vmatrix} a_{11} & c_1 \\ a_{21} & c_2 \end{vmatrix}}{\begin{vmatrix} a_{11} & a_{12} \\ a_{21} & a_{22} \end{vmatrix}}$$

PROCEDURES

<u>Procedure 1.</u> Although it is possible to address one-dimensional *subscripted* storage locations, as in A[1], A[2], etc., double-subscripted storage locations are not possible. So, for our purposes, it will be easier to use 4 separate locations (in the case of 2-by-2 determinants) or 9 (in the case of 3-by-3 determinants). Thus, think of the 2-by-2 determinant as having the components E, F, G, H:

$$\begin{vmatrix} a_{11} & a_{12} \\ a_{21} & a_{22} \end{vmatrix} \longrightarrow \begin{vmatrix} E & F \\ G & H \end{vmatrix}$$

and the 3-by-3 determinant as having the nine components O through W:

$$\begin{vmatrix} b_{11} & b_{12} & b_{13} \\ b_{21} & b_{22} & b_{23} \\ b_{31} & b_{32} & b_{33} \end{vmatrix} \longrightarrow \begin{vmatrix} O & P & Q \\ R & S & T \\ U & V & W \end{vmatrix}$$

The program will be entered as program #7. Recall, to write the program enter [MODE] 2 [EXE] and then choose 7 followed by [EXE].

Program # P7

Logic of what must be done:

1. Prompt for the *n* value and store it in register N.

2. If *n* equals 2, we will calculate a 2-by-2 determinant at label 2; if $n = 3$, we go to label 3 for the calculation.

3. Label 2. Here we enter the four elements of the matrix, calculate, and store the result in storage location A.

Casio syntax

1. "ENTER 2 OR 3" ? → N :

2. N = 2⇒ Goto 2: Goto 3:

3. Lbl 2 : "ENTER 1,1:"?→ E :
   "ENTER 1,2:"?→ F :
   "ENTER 2,1:"?→ G :
   "ENTER 2,2:"?→ H :
   (E × H − F × G)→ A ◄

4. Label 3. The other option is here. The nine elements of the matrix are entered and the determinant calculation is then stored in storage location B.

4. Lbl 3 : "ENTER 1,1:"?→ O :
"ENTER 1,2:"?→ P :
"ENTER 1,3:"?→ Q :
"ENTER 2,1:"?→ R :
"ENTER 2,2:"?→ S :
"ENTER 2,3:"?→ T :
"ENTER 3,1:"?→ U :
"ENTER 3,2:"?→ V :
"ENTER 3,3:"?→ W :
(OSW + PTU + QRV − QSU − OTV − PRW)→B ◄

<u>Procedure 2.</u> Use program # P7 to verify the calculations of the following determinants:

$$\begin{vmatrix} 2 & -3 \\ 1 & 2 \end{vmatrix} = 7 \qquad \begin{vmatrix} 0 & 2 & 1 \\ 3 & -1 & 2 \\ 4 & 0 & 1 \end{vmatrix} = 14$$

<u>Procedure 3.</u> Determinants can be applied to solving a system of linear equations using a procedure known as <u>Cramer's Rule</u>. Consider the system of 3 equations in 3 unknowns below:

$$x - y + 6z = 8$$
$$3x - y - 2z = 6$$
$$2x - 3y + 4z = 16$$

To solve for x, for example, we need to calculate two determinants and form a fraction with one determinant as the numerator and the other as the denominator:

$$x = \frac{\begin{vmatrix} 8 & -1 & 6 \\ 6 & -1 & -2 \\ 16 & -3 & 4 \end{vmatrix}}{\begin{vmatrix} 1 & -1 & 6 \\ 3 & -1 & -2 \\ 2 & -3 & 4 \end{vmatrix}}.$$ Using program # P7 we obtain: $x = \frac{-36}{-36} = 1$.

Similarly, we can find $y = \dfrac{\begin{vmatrix} 1 & 8 & 6 \\ 3 & 6 & -2 \\ 2 & 16 & 4 \end{vmatrix}}{\begin{vmatrix} 1 & -1 & 6 \\ 3 & -1 & -2 \\ 2 & -3 & 4 \end{vmatrix}} = \dfrac{144}{-36} = -4$ and

$$z = \dfrac{\begin{vmatrix} 1 & -1 & 8 \\ 3 & -1 & 6 \\ 2 & -3 & 16 \end{vmatrix}}{\begin{vmatrix} 1 & -1 & 6 \\ 3 & -1 & -2 \\ 2 & -3 & 4 \end{vmatrix}} = \dfrac{-18}{-36} = \dfrac{1}{2}.$$

Of course if the determinant of the coefficient matrix is zero, then the system will not have a unique solution; it will be either an inconsistent system or a dependent one. (See exercise 10.)

Name _____   Experiment #17

Date _____   **Exercise Sheet**

DIRECTIONS: Show all work to each question right on these pages.

In Exercises 1 through 6, use program #P7 to evaluate the given determinant.

1. $\begin{vmatrix} 5 & 6 \\ -2 & 7 \end{vmatrix}$ = _____   2. $\begin{vmatrix} 1 & 3 \\ 2 & 5 \end{vmatrix}$ = _____

3. $\begin{vmatrix} 3 & 1 & -7 \\ 0 & 2 & 4 \\ 8 & -1 & 6 \end{vmatrix}$ = _____   4. $\begin{vmatrix} 5 & 3 & 14 \\ 0 & 8 & -6 \\ 4 & 0 & 0 \end{vmatrix}$ = _____

5. $\begin{vmatrix} 1 & 2 & 3 \\ 4 & 5 & 6 \\ 7 & 8 & 9 \end{vmatrix}$ = _____   6. $\begin{vmatrix} 1 & 0 & 0 \\ 0 & 1 & 0 \\ 0 & 0 & 1 \end{vmatrix}$ = _____

In Exercises 7 through 9, use Cramer's Rule and the Casio to solve the system of equations.

7. $2x + 3y = 5$
   $6x - 2y = 4$

8. $x + y = \frac{1}{2}$
   $4x - 8y = -7$

x = _____ x = _____

y = _____ y = _____

9. $x + y + z = 11$ $x = \underline{\hspace{2cm}}$

    $x + y - z = -1$ $y = \underline{\hspace{2cm}}$

    $3x + 5y - 2z = 11$ $z = \underline{\hspace{2cm}}$

10. a. The following system is inconsistent. Verify this by using the Casio to graph the system (as in Experiment #16):

$$2x - 4y = 7$$
$$3x - 6y = 8$$

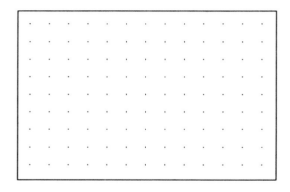

b. The following system is dependent:

$$2x - 4y = 6$$
$$3x - 6y = 9$$

Explain in your own words how determinants can be used to see if a system is inconsistent, dependent, or if a unique solution exists.

\underline{\hspace{14cm}}

\underline{\hspace{14cm}}

\underline{\hspace{14cm}}

Conic Sections Part I:
Circles and Parabolas

INTRODUCTION

1. The general formula for a circle of radius r centered at (h, k) is $(x-h)^2 + (y-k)^2 = r^2$. A circle is not a function and to graph it on the Casio (as we have already done in Experiment #16), we enter it as two (semi-circular) functions:

$$y = \sqrt{r^2 - (x-h)^2} + k \quad \text{and} \quad y = -\sqrt{r^2 - (x-h)^2} + k$$

2. Parabolas were first graphed as general quadratic functions in Experiment #4. Here, we present a variation in the form of the vertical axis parabola whose vertex is at (h, k) and has focal length of p:[17] $(x-h)^2 = 4p(y-k)$. Again, to adapt it for graphing y as a function of x for the Casio, we would write it as $y = \dfrac{(x-h)^2}{4p} + k$. [The parabola opens upward if $p > 0$ and downward if $p < 0$.]

3. The form for a general horizontal axis parabola whose vertex is (h, k) and which has focal length p is $(y-k)^2 = 4p(x-h)$. Since a horizontal axis parabola cannot be represented by one function, we again need to enter two functions for the Casio to handle its graph. They are: $y = \sqrt{4p(x-h)} + k$ and $y = -\sqrt{4p(x-h)} + k$. [The parabola opens to the right if $p > 0$ and to the left if $p < 0$.]

PROCEDURES

Procedure 1. One way to graph the circle $(x-2)^2 + (y+1)^2 = 2$ is to store 2, -1, and $\sqrt{2}$ in storage locations H, K, and R, respectively. Then, graph the two functions $y = \sqrt{r^2 - (x-h)^2} + k$ and $y = -\sqrt{r^2 - (x-h)^2} + k$. The keystrokes for this appear below:

---

[17]Saying a parabola has a focal length of p units means that the focus is p units from the vertex on the axis of symmetry. Equivalently, we could say that the directrix is the line $y = k - p$ (for vertical axis parabolas) or $x = h - p$ (for horizontal axis parabolas).

2 ⇨ ALPHA H EXE

(−) 1 ⇨ ALPHA K EXE

√ 2 ⇨ ALPHA R EXE

Graph √ ((ALPHA R x^2 − ((ALPHA X − ALPHA H)) x^2)) + ALPHA K EXE

Graph (−) ((√ ((ALPHA R x^2 − ((ALPHA X − ALPHA H)) x^2)))) + ALPHA K EXE

When entering your range settings, keep in mind that the domain of this relation is $h - r \leq x \leq h + r$ and the range is $k - r \leq y \leq k + r$. We graph $(x-2)^2 + (y+1)^2 = 2$ in Figure 18.1 below.

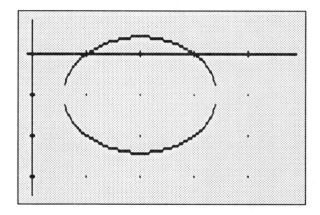

```
Range
Xmin:    − .1
 max:    4.5
 scl:    1.
Ymin:    −3.2
 max:    .5
 scl:    1.
```

The gap between the two semicircles occurs because of the distances between pixels. It is nothing to worry about.

FIGURE 18.1. $(x-2)^2 + (y+1)^2 = 2$, a circle of radius $\sqrt{2}$ centered at $(2, -1)$.

Actually, because of our choices in the range settings above, the "circle" looks more like a watermelon! This has to do with the ratio of X and Y distances on the Casio graphing screen. The screen is 95 columns of pixels (for x) by 63 rows of pixels (for y). To preserve that ratio (just about 3:2), the quantity $(\text{Xmax} - \text{Xmin}) \div (\text{Ymax} - \text{Ymin})$ would have to be 3:2. If that ratio is 3:2, the circle's shape (or *eccentricity*) would be preserved. To illustrate this, Figure 18.2 is the same graphing with a 3:2 ratio for x distance to y distance.

```
Range
Xmin:   -1.
 max:   4.5
 scl:   1.
Ymin:   -3.2
 max:   .5
 scl:   1.
```

Notice that the length of X above is 5.5 and the length of Y is 3.7. That is very close to a 3:2 ratio.

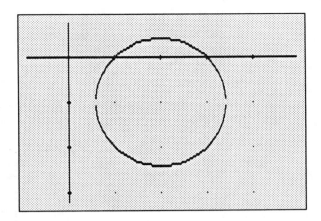

FIGURE 18.2. A rounder circle.

<u>Procedure 2.</u> We will graph the parabola whose vertex is at $(3, -2)$ and whose focal length is $\frac{1}{2}$. Thus, $h = 3$, $k = -2$, and $p = \frac{1}{2}$. We enter the information with the following keystrokes:

3 ⇨ ALPHA H EXE

(−) 2 ⇨ ALPHA K EXE

0.5 ⇨ ALPHA P EXE

Graph ((ALPHA X − ALPHA H)) x^2 ÷ ((4 × ALPHA P)) + ALPHA K EXE

```
Range
Xmin:   -4.
 max:   8.
 scl:   2.
Ymin:   -3.
 max:   10
 scl:   2.
```

Vertex: $(3, -2)$
Focus: $(3, -1.5)$
Directrix: $y = -2.5$

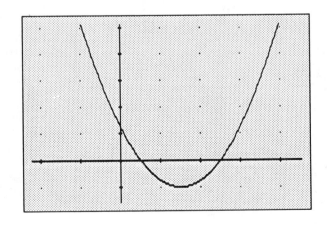

FIGURE 18.3. $(x-3)^2 = 2(y+2)$

In the exercises, you will be asked to generalize the above discussion and write a program to display the graph and the coordinates of the focus.

<u>Procedure 3.</u> A parabola whose axis of symmetry is a horizontal line can be graphed. We must break the equation up as two functions, like we did for the circle. For example, to graph $(y+1)^2 = -4(x-2)$, solve for y:

$$(y+1)^2 = -4(x-2)$$
$$y + 1 = \pm\sqrt{-4(x-2)}$$
$$y = \pm\sqrt{-4(x-2)} - 1$$

So, we graph $y = \sqrt{-4(x-2)} + 1$ and $y = -\sqrt{-4(x-2)} + 1$. Notice that h is 2, k is -1 and $p = -1$. The key sequence is shown below:

2 ⇨ ALPHA H EXE

(−) 1 ⇨ ALPHA K EXE

(−) 1 ⇨ ALPHA P EXE

Graph √ (((−) 4 × ((ALPHA X − ALPHA H)))) + ALPHA K EXE

Graph (−) √ (((−) 4 × ((ALPHA X − ALPHA H)))) + ALPHA K EXE

The parabola is graphed below in Figure 18.4.

Range	
Xmin:	−6.
max:	3.
scl:	2.
Ymin:	−10.
max:	6.
scl:	2.

Vertex: (2, -1)
Focus: (1, -1)
Directrix: $x = 3$

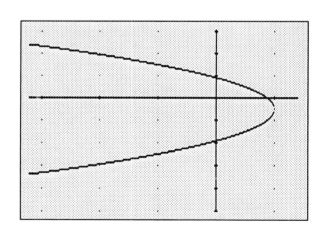

FIGURE 18.4. $(y+1)^2 = -4(x-2)$

Experiment #18
Page 122

DIRECTIONS: Show all work to each question right on these pages. For graphs, make a sketch based on the Casio's graphing screen display.

In questions 1 and 2, graph the circle represented by the given equation. Also, state the center and radius of the circle.

1. $(x-3)^2 + (y+2)^2 = 4$
   Center: _____
   Radius: _____

2. $(x+1)^2 + (y-1)^2 = 3$
   Center: _____
   Radius: _____

In questions 3 through 6, graph the parabola represented by the given equations. Also, state the parabola's vertex, focus, and directrix.

3. $y^2 = 2x$
   Vertex: _____
   Focus: _____
   Directrix: _____

4. $(x-2)^2 = 2(y+4)$
   Vertex: _____
   Focus: _____
   Directrix: _____

5. $x^2 = 3(y-3)$

    Vertex: _____

    Focus: _____

    Directrix: _____

6. $(y+2)^2 = -6(y+4)$

    Vertex: _____

    Focus: _____

    Directrix: _____

7. Write a program to run on your Casio that will accept for input the values of h, k, and p and then graph the vertical axis parabola and display its focus. List the program steps on a blank piece of paper and attach it to this page.

In questions 8 through 11, each of the parabolas was graphed on a Casio with the following range settings: Xmin:-10, Xmax:10, Xscl:2, Ymin:-8, Ymax:8, Yscl:2. State the *equation* that represents each parabola. [Hint: all vertices are at (1, -2) and p is ± 1 or ± 2 in each case.]

8. _____

9. _____

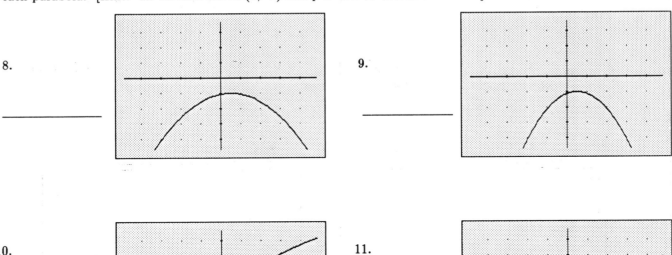

10. _____

11. _____

Conic Sections Part II:
Ellipses

INTRODUCTION

1. An <u>ellipse</u> is the set of all points, the sum of whose distances from two distinct fixed points (called <u>foci</u>) is a constant. That constant has the value $2a$. If the major axis of the ellipse is horizontal, the equation of the ellipse centered at (h, k) is:

$$\frac{(x-h)^2}{a^2} + \frac{(y-k)^2}{b^2} = 1$$

2. If the major axis is vertical, the equation of the ellipse is:

$$\frac{(x-h)^2}{b^2} + \frac{(y-k)^2}{a^2} = 1$$

3. In either case above, a represents half the length of the major axis and b represents half the length of the minor axis. The distance a focus point is from the ellipse's center is c units and the relationship among a, b, and c is $c^2 = a^2 - b^2$.

4. The <u>eccentricity</u> of an ellipse is $\frac{c}{a}$. It is a measure of the ellipse's "roundness"; if the eccentricity is near zero, the ellipse is nearly circular and as the eccentricity approaches 1, the ellipse becomes more elongated.

PROCEDURES

<u>Procedure 1.</u> As with circles, ellipses are not functions and in order to graph them on the Casio, we must graph two semi-ellipses. In the case of a horizontal major axis ellipse, we solve for y and get the following two functions of x:

$$y = \frac{b}{a}\sqrt{a^2 - (x-h)^2} + k \qquad\qquad y = -\frac{b}{a}\sqrt{a^2 - (x-h)^2} + k$$

For example, consider

$$\frac{(x-2)^2}{9} + \frac{(y-1)^2}{5} = 1$$

Here, $h = 2$, $k = 1$, $a = 3$, $b = \sqrt{5}$ and $c = 2$. The keystrokes needed in order to obtain the graph in Figure 19.1 appear below.

2 ⇨ [ALPHA] [H] [EXE]

1 ⇨ [ALPHA] [K] [EXE]

3 ⇨ [ALPHA] [A] [EXE]

[√] 5 ⇨ [ALPHA] [B] [EXE]

[Graph] [(] [ALPHA] [B] [÷] [ALPHA] [A] [)] [√] [(] [ALPHA] [A] [x^2] [−] [(] [ALPHA] [X] [−] [ALPHA] [H] [)] [x^2] [)] [+] [ALPHA] [K] [EXE]

[Graph] [(] [(−)] [ALPHA] [B] [÷] [ALPHA] [A] [)] [√] [(] [ALPHA] [A] [x^2] [−] [(] [ALPHA] [X] [−] [ALPHA] [H] [)] [x^2] [)] [+] [ALPHA] [K] [EXE]

Range	
Xmin:	-1.4
max:	5.4
scl:	$1.$
Ymin:	-1.25
max:	3.24
scl:	$1.$

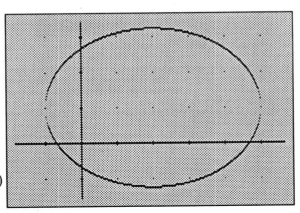

To correctly represent the ellipse's shape on the Casio graphing screen, make sure that the ratio (Xmax-Xmin):(Ymax-Ymin) approximates 3:2, the pixel ratio for the graphing screen.

FIGURE 19.1. $\dfrac{(x-2)^2}{9} + \dfrac{(y-1)^2}{5} = 1.$

<u>Procedure 2.</u> In Figure 19.1, the ellipse's eccentricity is $\frac{c}{a} = \frac{2}{3}$. In the next example, we want to graph $\frac{(x-1)^2}{4} + \frac{(y+2)^2}{16} = 1$ where $h = 1$, $k = -2$, $a = 4$, $b = 2$, and $c = 2\sqrt{3}$. The ellipse in Figure 19.2 has eccentricity $\frac{\sqrt{3}}{2} \approx 0.866$ and is more elongated than the ellipse in Figure 19.2. To have the Casio enter the graph for a vertical axis ellipse, we have to solve the general vertical axis ellipse for y:

$$y = \pm \frac{a}{b}\sqrt{b^2 - (x-h)^2} + k$$

The keystrokes to enter the two functions are:

1 ➡ |ALPHA| |H| |EXE|

|(−)| 2 ➡ |ALPHA| |K| |EXE|

4 ➡ |ALPHA| |A| |EXE|

2 ➡ |ALPHA| |B| |EXE|

|Graph| |(| |ALPHA| |A| |÷| |ALPHA| |B| |)| |√| |(| |ALPHA| |B| |x^2| |−| |(| |ALPHA| |X| |−| |ALPHA| |H| |)| |x^2| |)| |+| |ALPHA| |K| |EXE|

|Graph| |(| |(−)ALPHA| |A| |÷| |ALPHA| |B| |)| |√| |(| |ALPHA| |B| |x^2| |−| |(| |ALPHA| |X| |−| |ALPHA| |H| |)| |x^2| |)| |+| |ALPHA| |K| |EXE|

Range
Xmin: $-5.$
 max: 7.
 scl: 1.
Ymin: $-6.$
 max: 2.
 scl: 1.

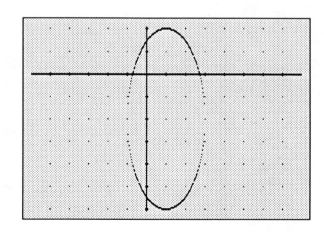

Vertices: $(1,2)$ and $(1, -6)$

Foci: $(1, -2 \pm 2\sqrt{3})$

Eccentricity: $\frac{\sqrt{3}}{2} \approx 0.866$

FIGURE 19.2. $\frac{(x-1)^2}{4} + \frac{(y+2)^2}{16} = 1$

We conclude this experiment with a summary of the information about ellipses.

For Horizontal Major Axis Ellipses:

The general formula is $\dfrac{(x-h)^2}{a^2} + \dfrac{(y-k)^2}{b^2} = 1$

where $2a$ is the length of the major axis and $2b$ is the length of the minor axis. Also, the vertices are at $(h \pm a,\ k)$ and the foci are at $(h \pm c,\ k)$ where $c = \sqrt{a^2 - b^2}$.

For Vertical Major Axis Ellipses:

The general formula is $\dfrac{(x-h)^2}{b^2} + \dfrac{(y-k)^2}{a^2} = 1$

where $2a$ is the length of the major axis and $2b$ is the length of the minor axis. Also, the vertices are at $(h,\ k \pm a)$ and the foci are at $(h,\ k \pm c)$ where $c = \sqrt{a^2 - b^2}$.

DIRECTIONS: Show all work to each question right on these pages. For graphs, make a sketch based on the Casio's graphing screen display.

1. Refer to Figure 19.1. What are the coordinates of the vertices and foci? What is the domain and range of this relation?

Coordinates of vertices: _____   Foci: _____

Domain: _____Range: _____

2. What is the domain and range of the relation graphed in Figure 19.2?

Domain: _____Range: _____

In Exercises 3 through 6, graph the ellipse whose equation is given. In each case state the coordinates of the center, vertices and foci and find the ellipses's eccentricity.

3. $\frac{x^2}{25} + \frac{y^2}{16} = 1$

    Center: _____

    Vertices: _____

    Foci: _____

    Eccentricity: _____

4. $\dfrac{x^2}{16} + \dfrac{y^2}{25} = 1$

   Center: _____

   Vertices: _____

   Foci: _____

   Eccentricity: _____

5. $\dfrac{(x-1)^2}{9} + \dfrac{(y+5)^2}{16} = 1$

   Center: _____

   Vertices: _____

   Foci: _____

   Eccentricity: _____

6. $\dfrac{(x-4)^2}{9} + y^2 = 1$

   Center: _____

   Vertices: _____

   Foci: _____

   Eccentricity: _____

Conic Sections Part III:
Hyperbolas

INTRODUCTION

1. An <u>hyperbola</u> is the set of all points, the absolute value of the difference of whose distances from two fixed points (called <u>foci</u>) is constant. The line segment connecting two vertices of the hyperbola is called the <u>transverse axis</u>. If the transverse axis is horizontal, the general form for an hyperbola centered at (h, k) is:

$$\frac{(x-h)^2}{a^2} - \frac{(y-k)^2}{b^2} = 1$$

2. If the transverse axis is vertical, the equation is:

$$\frac{(y-k)^2}{a^2} - \frac{(x-h)^2}{b^2} = 1$$

In either case, the relationship among a, b, and c is $c^2 = a^2 + b^2$. The foci are c units from the center and the vertices are a units from the center.

3. Every hyperbola has two separate branches. Also, these branches approach straight lines asymptotically. The equations of the asymptotes are:

$$y = \pm \frac{b}{a}(x-h) + k \qquad \text{(for horizontal transverse axis hyperbolas)}$$
$$\text{or}$$
$$y = \pm \frac{a}{b}(x-h) + k \qquad \text{(for vertical transverse axis hyperbolas)}$$

PROCEDURES

<u>Procedure 1.</u> We begin with examining the graph of $\frac{y^2}{25} - \frac{x^2}{36} = 1$, a vertical transverse axis hyperbola centered at the origin with $a = 5$, $b = 6$ and $c = \sqrt{61}$. Again, hyperbolas are not

functions and we must solve for y and graph the hyperbola as two separate functions. In general, we have

$$y = \pm\frac{b}{a}\sqrt{(x-h)^2 - a^2} + k \qquad \text{(for horizontal transverse axis hyperbolas)}$$

or

$$y = \pm\frac{a}{b}\sqrt{(x-h)^2 + b^2} + k \qquad \text{(for vertical transverse axis hyperbolas)}$$

The keystrokes for graphing $\dfrac{y^2}{25} - \dfrac{x^2}{36} = 1$

0 ⇒ ALPHA H EXE

0 ⇒ ALPHA K EXE

5 ⇒ ALPHA A EXE

6 ⇒ ALPHA B EXE

Graph ((ALPHA A ÷ ALPHA B)) √ ((((ALPHA X − ALPHA H)) x^2 + ALPHA B x^2)) + ALPHA K EXE

Graph (((−) ALPHA A ÷ ALPHA B)) √ ((((ALPHA X − ALPHA H)) x^2 + ALPHA B x^2)) + ALPHA K EXE

In addition, to graph the asymptotes on the same axes, we should add:

Graph ((ALPHA A ÷ ALPHA B)) ((ALPHA X − ALPHA H)) + ALPHA K

Graph (((−) ALPHA A ÷ ALPHA B)) ((ALPHA X − ALPHA H)) + ALPHA K

The graph of $\dfrac{y^2}{25} - \dfrac{x^2}{36} = 1$ appears in Figure 20.1.

```
Range
Xmin:   -15.
  max:   15.
  scl:   3.
Ymin:   -10.
  max:   10.
  scl:   3.
```

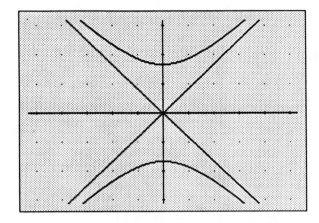

The asymptotes are $y = \frac{5}{6}x$ and $y = -\frac{5}{6}x$.

FIGURE 20.1. $\dfrac{y^2}{25} - \dfrac{x^2}{36} = 1$ and $y = \pm\dfrac{5}{6}x$.

Notice that the domain of this relation is **R** and its range is $y \le -5$ or $y \ge 5$. Also, the foci occur at $(h, k+c)$ and at $(h, k-c)$ or at $(0, \sqrt{61})$ and $(0, -\sqrt{61})$

<u>Procedure 2.</u> Finally, we graph the horizontal transverse axis hyperbola $\dfrac{(x+1)^2}{144} - \dfrac{(y-4)^2}{25} = 1$ in Figure 20.2. The keystrokes entered to obtain that graph are:

$\boxed{(-)}$ 1 $\boxed{\rightarrow}$ $\boxed{\text{ALPHA}}$ $\boxed{\text{H}}$ $\boxed{\text{EXE}}$

4 $\boxed{\rightarrow}$ $\boxed{\text{ALPHA}}$ $\boxed{\text{K}}$ $\boxed{\text{EXE}}$

12 $\boxed{\rightarrow}$ $\boxed{\text{ALPHA}}$ $\boxed{\text{A}}$ $\boxed{\text{EXE}}$

5 $\boxed{\rightarrow}$ $\boxed{\text{ALPHA}}$ $\boxed{\text{B}}$ $\boxed{\text{EXE}}$

$\boxed{\text{Graph}}$ $\boxed{(}$ $\boxed{\text{ALPHA}}$ $\boxed{\text{B}}$ $\boxed{\div}$ $\boxed{\text{ALPHA}}$ $\boxed{\text{A}}$ $\boxed{)}$ $\boxed{\sqrt{\ }}$ $\boxed{(}$ $\boxed{(}$ $\boxed{\text{ALPHA}}$ $\boxed{\text{X}}$ $\boxed{-}$ $\boxed{\text{ALPHA}}$ $\boxed{\text{H}}$ $\boxed{)}$ $\boxed{x^2}$ $\boxed{-}$ $\boxed{\text{ALPHA}}$ $\boxed{\text{A}}$ $\boxed{x^2}$ $\boxed{)}$ $\boxed{+}$ $\boxed{\text{ALPHA}}$ $\boxed{\text{K}}$ $\boxed{\text{EXE}}$

$\boxed{\text{Graph}}$ $\boxed{(}$ $\boxed{(-)}$ $\boxed{\text{ALPHA}}$ $\boxed{\text{B}}$ $\boxed{\div}$ $\boxed{\text{ALPHA}}$ $\boxed{\text{A}}$ $\boxed{)}$ $\boxed{\sqrt{\ }}$ $\boxed{(}$ $\boxed{(}$ $\boxed{\text{ALPHA}}$ $\boxed{\text{X}}$ $\boxed{-}$ $\boxed{\text{ALPHA}}$ $\boxed{\text{H}}$ $\boxed{)}$ $\boxed{x^2}$ $\boxed{-}$ $\boxed{\text{ALPHA}}$ $\boxed{\text{A}}$ $\boxed{x^2}$ $\boxed{)}$ $\boxed{+}$ $\boxed{\text{ALPHA}}$ $\boxed{\text{K}}$ $\boxed{\text{EXE}}$

```
Range
Xmin:   -24.
 max:    21.
 scl:     3.
Ymin:   -15.
 max:    15.
 scl:     3.
```

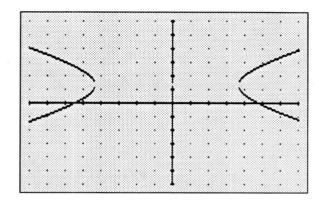

Domain: $x \leq -13$ or $x \geq 11$

Range: **R**

FIGURE 20.2. $\dfrac{(x+1)^2}{144} - \dfrac{(y-4)^2}{25} = 1$

The asymptotes were not drawn but in exercise 2 you are asked to calculate them and use them to approximate points on the hyperbola.

DIRECTIONS: Show all work to each question right on these pages. For graphs, make a sketch based on the Casio's graphing screen display.

1. A student was asked to find the value of y when $x = 60,000$ for the hyperbola $\dfrac{y^2}{25} - \dfrac{x^2}{36} = 1$ graphed in Figure 20.1. Without a calculator her answer was "about $\pm 50,000$". How did she come up with the answer so quickly?

2. a) What are the equations of the asymptotes for the hyperbola graphed in Figure 20.2?

   Asymptotes: _____

b) Use your answer to part a to approximate the value of y when x is 11,999.

   $y =$ _____

In questions 3 through 6, graph the hyperbola with the given equation. Also, state the domain and range of the hyperbola, its vertices and foci coordinates, and the equations of its asymptotes. Include the asymptotes on the graph.

3. $\dfrac{(x-2)^2}{4} - \dfrac{(y-2)^2}{5} = 1$

Vertices: _____

Foci: _____

Asymptotes: _____

Domain: _____

Range: _____

4. $\dfrac{(y-2)^2}{4} - \dfrac{(x-2)^2}{5} = 1$

Vertices: _____

Foci: _____

Asymptotes: _____

Domain: _____

Range: _____

5. $\dfrac{(x+3)^2}{16} - (y-2)^2 = 1$

Vertices: _____

Foci: _____

Asymptotes: _____

Domain: _____

Range: _____

6. $\dfrac{(y-1)^2}{\frac{1}{4}} - \dfrac{(x+3)^2}{\frac{1}{9}} = 1$

Vertices: _____

Foci: _____

Asymptotes: _____

Domain: _____

Range: _____

Polar Coordinates

INTRODUCTION

1. A point P in the plane has unique rectangular coordinate system coordinates (x, y). The point can also be labeled in the <u>polar coordinate system</u> with (r, θ) where r is the directed distance from the origin to point P and θ is the directed angle, measured counterclockwise, through P. The polar representation of a point is *not* unique.

2. The relationships among x, y, r, and θ are given by the following four equations:

$$x = r\cos\theta \qquad\qquad y = r\sin\theta$$
$$\tan\theta = \frac{y}{x} \qquad\qquad r^2 = x^2 + y^2$$

3. It is best to be in radian mode ($\boxed{\text{MODE}}$ 5) for the entirety of this experiment.

PROCEDURES

<u>Procedure 1.</u> We begin by using the Casio calculator to do conversions from rectangular to polar and from polar to rectangular coordinates. The two keys that do these conversions are the $\boxed{\text{SHIFT}}$ $\boxed{\text{Pol}(x,y)}$ and the $\boxed{\text{SHIFT}}$ $\boxed{\text{Rec}(r,\theta)}$ keys (found on the $\boxed{+}$ and the $\boxed{-}$ keys).

To convert $(3, 4)$ to polar, perform the following keystrokes:

$\boxed{\text{SHIFT}}$ $\boxed{\text{Pol}(x,y)}$ 3 $\boxed{\text{SHIFT}}$ $\boxed{,}$ 4 $\boxed{)}$ $\boxed{\text{EXE}}$

You should see a 5 displayed as the value of r. Also, this value is automatically stored in memory storage location I. The value of θ is stored in location J. So, by pressing $\boxed{\text{ALPHA}}$ $\boxed{\text{J}}$ $\boxed{\text{EXE}}$ we see θ (in radians) is 0.92730. So, $(3, 4)$ in rectangular coordinates is represented as $(5, 0.9273)$ in polar coordinates.

Now try to convert the rectangular $(-5, -12)$ to polar:

SHIFT Pol(x,y) (−) 5 SHIFT , (−) 12) EXE

The displayed value of r is 13; θ is returned by pressing ALPHA J EXE. It is -1.96559.

To convert from the polar coordinates of $(2.3, -\frac{\pi}{8})$, we enter the following:

SHIFT Rec(r,θ) 2.3 SHIFT , (−) SHIFT π ÷ 8) EXE

The value of x, 2.12492, is displayed (and, again storage location I is used); the y value of the rectangular coordinates is -0.88017 and is stored in J.

Procedure 2. To plot a polar curve, such as $r = 1 - 2\cos\theta$, we will employ the following program. Store the program as program # P9.

Program # P9

Logic of what must be done:

1. Establish appropriate range settings for the Casio. Keep in mind we will want to preserve the 3:2 pixel ratio for x:y. Alternatively, we could ask for the values to be input upon program execution.

2. Think of θ at T. We must initialize T at 0 and let it increment to 2π in small increments. (See steps 6 and 7 also.)

3. Calculate R for any given T value.

4. Do the conversion from polar to rectangular coordinates. Remember, the Casio graphing screen has to have rectangular coordinates to PLOT!

5. Now the value of x that needs to be plotted -- it is stored in I -- along with

Casio syntax

1. Range -4, 2, 1, -2, 2, 1:

2. 0 → T :

3. Lbl 1: $1 - 2\cos T$ → R:

4. Rec(R,T) :

5. Plot I, J :

y (stored in J).

6. Increment T here. We use 0.05. The smaller incrementer, the closer together the plotted points will be but the longer it will take to complete the plot.

6. T + 0.05→T :

7. Continue the Label 1 loop until the value of T increases to 2π.

7. T < 2π ⇒ Goto 1 :

8. Recall this "trick" to leave the program in graphing screen mode so that the polar plot is visible at the program's end. (Without this statement, you would need to hit the $\boxed{G \leftrightarrow T}$ key at program's end to view the graph.)

8. Graph y = 0 ◄

The result of running this program appears in Figure 21.1. The graph is called a limaçon and variations of it are examined in the exercises.

Range	
Xmin:	−4.
max:	2.
scl:	1.
Ymin:	−2.
max:	2.
scl:	1.

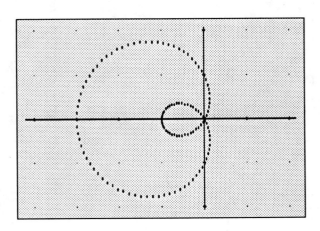

About 125 points get plotted using program P9 above. An alternative is to connect the points using the Line key. (A LINE statement would have to be inserted between statements 5 and 6).

FIGURE 21.1. $r = 1 - 2\cos\theta$, a limaçon.

Procedure 3.   Next, we will graph $r = 2\sin 5\theta$, a member of the rose curve family, a family that is also examined more fully in the exercises.

```
Range
Xmin:   -3.
 max:    3.
 scl:    1.
Ymin:   -2.
 max:    2.
 scl:    1.
```

Program P9 was altered to reflect the new function and the range settings above.

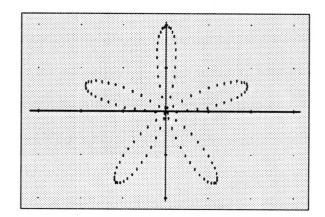

FIGURE 21.2. $r = 2\sin 5\theta$, a rose curve with five petals.

Procedure 4.   Finally, consider the polar curve, $r = 1 - \sin\theta$. This curve falls into the family of curves known as cardioids and is graphed below. Again, the range settings were changed from the original program P9 above.

```
Range
Xmin:   -3.
 max:    3.
 scl:    1.
Ymin:   -3.
 max:    1.
 scl:    1.
```

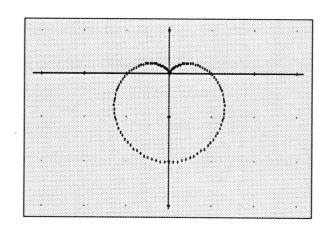

FIGURE 21.3.The cardioid $r = 1 - \sin\theta$.

Name _____   **Experiment #21**

Date _____   **Exercise Sheet**

DIRECTIONS: Show all work to each question right on these pages. For graphs, make a sketch based on the Casio's graphing screen display.

In questions 1 through 6, convert the coordinates of the point from one system to the other by completing the table.

	Rectangular Coordinates	Polar Coordinates
1.	$(-3, 4)$	_____
2.	_____	$(3, \frac{3\pi}{2})$
3.	$(-8, 15)$	_____
4.	$(-2.45, 6.91)$	_____
5.	_____	$(2.5, 4.93)$
6.	_____	$(6.5, -2.54)$

A polar curve of the form $r = a \cos n\theta$ or $r = a \sin n\theta$ $(n \geq 2)$ is called a <u>rose curve</u>. If the value of n is odd, the curve has n petals; if n is even, the curve has $2n$ petals. In questions 7 through 10, graph each curve on the polar coordinate system. [Adapt program #P9 to have the Casio plot the curve.]

7. $r = 2 \cos 3\theta$

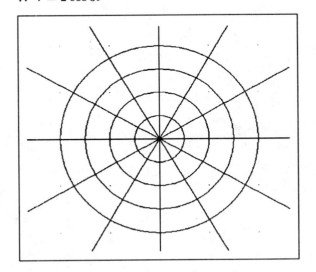

8. $r = 3 \sin 5\theta$

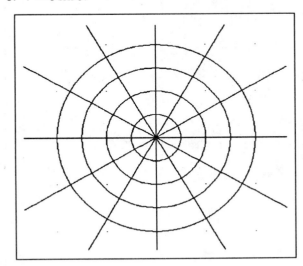

9. $r = 2\sin 2\theta$

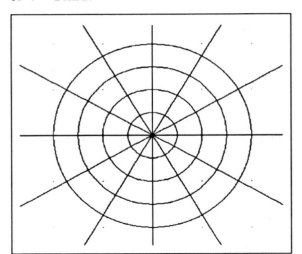

10. $r = 3\cos 4\theta$

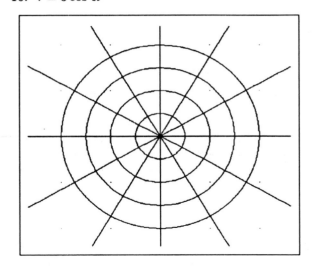

The family of polar curves called *limaçons* take the form $r = a \pm b\cos\theta$ or $r = a \pm b\sin\theta$ (where $a > 0$ and $b > 0$). In questions 11 and 12, sketch each curve.

11. $r = 1 + 2\sin\theta$ (A limaçon with inner loop)

12. $r = 3 + 2\cos\theta$ (A limaçon with dimple)

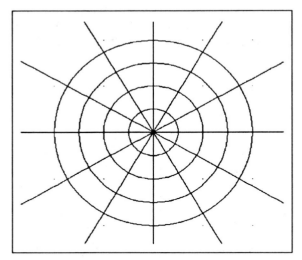

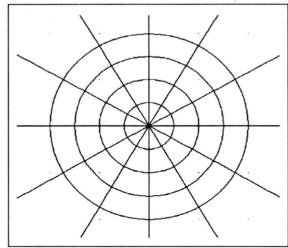

Casio Key Glossary

Use the diagrams on pages 149 and 150 together with the definitions and descriptions on the next few pages as references for using your calculator. Where appropriate, we have included page references for you to learn more detail about the key.

Key Reference or Category	Way Key is Depicted in this Manual	Description
All clear key	AC	Use to turn calculator back on if automatic shut off occurs. Used mainly to clear text display. See pp. 9, 14, 35, 147.
Alphabet key	ALPHA	Use in conjunction with the alphabet and related characters. An **A** will appear on the screen until the alphabetic character is entered. If preceded by the SHIFT key, calculator in a shift-lock mode. See pp. 2, 4-6.
Answer key	Ans ALPHA SPACE	Recalls the last computation. See page 4. Use to enter a space.
Assignment key	→	Use to assign a value to a memory. See pp. 12-15, 28, 34, 69, 114.
Cursor/Replay keys[19]	△ ◁ ▷ ▽	Use to move the cursor up, left, right, or down for editing an expression. After a command has been executed, use ◁ or ▷ to "replay" that command. The TRACE command works in conjunction with ◁ and ▷ for highlighting points on graphs. See pp. 3, 9, 20, 41, 147.
Delete key	DEL	Use to delete the character at the current cursor position. See pp. 5, 19.
	SHIFT Mcl	Use to delete memory contents or, when the range screen is displayed, to cause default range settings to appear. See pp. 5, 16.
Execute key	EXE	Use as an "enter" key for computation or data entry.

---

[19]These are the curosr movement keys on the *fx-7500G*. The *fx-7000G*'s cursor keys have a slightly different appearance and the right movement key, for example, looks like →.

Function keys	$\sqrt{}$	Use before an entry to display square root.
	x^2	Use after an entry to display square of number.
	$\log$	Use before an entry to display the common logarithm of number.
	$\ln$	Use before an entry to display the logarithm to the base e (natural log).
	x^{-1}	Use after an entry to display the reciprocal of number.
	10^x	Use before an entry to find its common antilogarithm.
	e^x	Use before an entry to raise it to the power of e (natural antilogarithm).
	$\sin$	Use before an entry to display the sine of that entry.
	SHIFT $\sin^{-1}$	Use before an entry to find the inverse sine.
	$\cos$	Use before an entry to display the cosine of that entry.
	SHIFT $\cos^{-1}$	Use before an entry to find the inverse cosine.
	$\tan$	Use before an entry to display the tangent of that entry.
	SHIFT $\tan^{-1}$	Use before an entry to find the inverse tangent.
	x^y	Use between two entries to raise the first entry to the second entry power.
	SHIFT Abs	Use before an entry to calculate its absolute value.
	$(-)$	Use before an entry to negate that entry.
	$\sqrt[x]{}$	Use between two entries to find the xth (first entry) root of the second entry.
	SHIFT $\sqrt[3]{}$	Use before an entry to find the cube root of that entry.

Graph keys	G↔T	Graph/text toggle. See pp. 15, 35, 139.
	Graph	Displays "y = " on screen then user must enter function to be graphed. See pp. 7, 12, 14, 20, 24, 33, 61, 83, 85, 91, 92, 110, 120, 126, 132, 133.
	Plot	Use to plot a single point on the graph screen. See pp. 14, 35, 147.
	Trace	Use to highlight previously drawn graph point by point. Either x or y coordinate is displayed. See pp. 19, 20, 36, 41, 62, 101, 108.
	Range	Use to view or set range values. See pp. 5 − 8, 12, 19, 20, 21, 83, 87.
	Line	For line graphs and regression lines.
	Cls	Use for clearing the graph display. See pp. 8, 13, 17, 83.
	SHIFT X↔Y	Alternates between x and y coordinate display for TRACE command.
Input key	?	Use in programs to allow for numerical entry during program execution. See pp. 13-15, 28.
Insert key	INS	Changes cursor from a flashing underline to a flashing box so that items can be inserted.
Mode key	MODE	Used to establish the type of angular measurement (degree or radian) and the way numbers are displayed. Some examples follow. See pp. 2, 5, 9, 11, 13, 14, 53

	1	Computation and program execution mode.
	2	Program editing and writing mode.
	3	Program clear (erase) mode.
	4	Degree mode.
	5	Radian mode.
	6	Grad (angular measurement) mode.
	7 n	To display n fixed number of decimal places.
	8 n	To display scientific notation with n significant digits. n must be 0 through 9.

Mode display key	M Disp	Use to display the system mode. System info is displayed only when the key is held down. See pp. 2, 67

Multiple statement key $\boxed{:}$ Use to combine two formulas for consecutive
                                                           computation or to separate multiple statements in
                                                           a program. See pp. 6, 12.

Program key $\boxed{\text{Prog}}\ \boxed{n}$ Executes program number n.

                       $\boxed{\text{SHIFT}}\ \boxed{\text{Goto}}$ Unconditional goto (jump) command.

Shift key $\boxed{\text{SHIFT}}$ Used in conjunction with gold functions on the
                                                           fx-7000G (blue functions on the fx-7500G). An
                                                           $\mathbf{S}$ will appear on the screen until the function key is
                                                           pressed. See pp. 2, 4.